ROBERT D. ALLEN

VICTOR VALLEY COLLEGE

CRITICAL THINKING CASE STUDY WORKBOOK

Second Edition

WCB McGraw-Hill

Boston, Massachusetts Burr Ridge, Illinois Dubuque, Iowa
Madison, Wisconsin New York, New York San Francisco, California St. Louis, Missouri

WCB/McGraw-Hill

A Division of The McGraw-Hill Companies

ISBN 0–697–34250–6

Printed in the United States of America

6 7 8 9 0 QPD 0 9 8 7 6 5 4 3 2 1

Contents

To the Student

How are the "facts" of biology discovered? What kinds of questions do biologists ask when doing experiments? How are experiments used to establish "facts"?

The case studies in this book demonstrate how experimental results and observations are used to establish the "facts" of biology. Studying these case studies gives you practice with analytical skills and with forming and testing hypotheses, making predictions, and critically evaluating results, as well as learning more about biological concepts.

These analytical procedures are termed *critical thinking skills.* Your ability to apply such skills will have a major impact on your understanding of biology.

Each case study provides introductory information about a specific topic and then forms a hypothesis that is investigated by a series of predictions and experimental investigations. The following diagram illustrates this procedure:

(Questions)
↓
Hypothesis → Prediction 1, Prediction 2, Prediction 3

Prediction 1	Prediction 2	Prediction 3
↓	↓	↓
Result 1	Result 2	Result 3

When a hypothesis is proposed and tested in this way, the following points should be kept in mind:

1. Although the hypothesis is sometimes introduced by a question, the question may not always be obvious. Sometimes the hypothesis is proposed as a statement of a cause and effect relationship without indicating a question.
2. *Several* predictions can be made based on the hypothesis. Each of these predictions can be experimentally tested and the results can be analyzed. The obvious questions are: Are the results as predicted? Do the results support or contradict the hypothesis?
3. Results may support, contradict, or neither support nor contradict the hypothesis. Only by carefully analyzing a result and looking for other interpretations can a decision be made about the strength of support or nonsupport for the hypothesis.
4. In many cases, subsequent predictions are based on the results of previous predictions. These subsequent predictions are often made to test or evaluate possible *alternative* interpretations of previous results. The analysis of alternative interpretations and the proposal of additional predictions to test these interpretations is illustrated in the case studies and should be recognized as an appropriate investigative procedure.
5. A hypothesis can *never* be proven with certainty. Instead, results may support or contradict the hypothesis. Thus, it is very important to test the hypothesis with as many different predictions and tests as possible. The more tests that support the hypothesis, the greater the confidence in the "truth" of the hypothesis. Thus, a "fact" is a well-supported hypothesis.

Keep these points in mind as you read the case studies:

- Look for other, alternative interpretations of the results.
- Think of other predictions that will test the hypothesis.
- Support for a hypothesis becomes stronger as *more* predictions and tests are done.

Questions

Questions are indicated with boxes and are repeated in a student answer sheet at the end of each case study. Make sure your answers are complete and that you clearly address the question. These questions are designed to test your understanding of the experimental procedures, analysis of the results, and biological concepts. Spend some time on these questions. Do not try to answer them too quickly. If you thoroughly think through and review the case study material to construct your answers, you will be rewarded with a much better understanding.

Levels of Difficulty

The case studies are written at 3 levels of difficulty:

Introductory (1)

These case studies are written for students with little experience practicing critical thinking skills and analyzing experimental procedures and biological concepts. In your first attempts with the case studies, try a few at the introductory level to test your ability. Several questions are raised in the study, and answers are provided to serve as examples. Other questions in the boxed sections are left for you to answer.

Intermediate (2)

As your skill improves, you will be able to attempt the intermediate level case studies with some success. The questions are more difficult than in the introductory level, and more experiments and observations are included.

Advanced (3)

As the most difficult level of case studies, these are challenging even for the best students. The questions are very difficult and require considerable thought.

You will probably find that you need to refer to your biology text for additional information on case study topics. The case studies do not attempt to provide detailed information on each topic, and you should use your text as necessary.

Do not underestimate the difficulty of these case studies. If you are like most students, you will find that these exercises offer a different view of science, one that requires some practice and thoughtful reflection. Do not be discouraged if, initially, you find these cases difficult.

Start with the introductory level and proceed to the higher levels as your skill develops. Your efforts to learn to think critically will result in a new appreciation of science.

PART I

Cell Physiology and Genetics

1 Case Study: The Lysosome (Difficulty Level 1)

Living cells carry out a wide variety of reactions and contain a great assortment of organic molecules and cellular structures. Among these molecular constituents are a number of digestive enzymes that can catalyze the breakdown of almost all of the molecules in the cell. Why do the enzymes *not* digest the other molecules in the cell? How can digestive enzymes and the molecules that the enzymes digest exist in the same cell? Are the enzymes isolated in some way so that the digestive enzymes and other cellular constituents do not come into contact?

Investigations to answer these questions involve breaking or grinding up the cell to release the enzymes present inside the cell and in the cell organelles. When the cell is completely broken up using an instrument such as a blender, all parts and organelles are disrupted. The material resulting from this treatment is termed the *cell homogenate.* Relatively high digestive enzyme activities (a measure of how rapidly products are produced in enzyme catalyzed reactions) are observed in the homogenate.

When less drastic procedures are used to disrupt the cells, however, enzyme activities are only about one-tenth as high as in the homogenate. In this treatment, most of the cell organelles are *not* disrupted. These results indicate that more gentle disruption procedures do not release most of the digestive enzymes and suggest the following hypothesis:

Hypothesis—Digestive enzymes inside cells are isolated in specialized organelles that usually prevent contact with digestible cell molecules. Under some conditions, however, the enzymes are released and utilized to perform their digestive function.

What organelles would contain and isolate the enzymes? Under what conditions are the enzymes released?

The first step in testing this hypothesis is to try to isolate the proposed organelle or organelles containing the enzymes. These organelles can be separated by centrifuging the cell homogenate at increasing speeds in a series of steps. The homogenate is placed in a centrifuge tube and spun at a low speed. The heaviest organelles and particles are forced to the bottom of the tube. The supernatant (i.e., the fluid left after the centrifugation) is removed and spun at a higher speed, which causes additional and heavier organelles to be forced to the bottom of the tube. By repeating this procedure, the components of the cell are separated. In one of the fractions, high digestive enzyme activities were measured, but any organelles present were too small to be observed with the compound light microscope. With the electron microscope, however, particles were observed that were surrounded by a membrane.

Further investigations using electron micrographs of cells indicated particles of similar appearance (fig. 1.1). A similar appearance between the isolated particles and those observed inside the cells is not enough evidence to conclude the particles are the same. Organelles could appear similar but have very different biochemical composition and function. To designate these particles as specific organelles in the cellular cytoplasm, they were termed *lysosomes.* Were these the structures containing the digestive enzymes? If so, what function did they serve?

Clues to answering these questions have been revealed by studies of white blood cells. These cells circulate in the bloodstream and engulf foreign material such as bacteria. After entering the cell, the bacterium is surrounded by a portion of the white cell's plasma membrane, which pinches off and isolates the bacterium from the cell interior as shown in figure 1.2. The bacterium is subsequently digested.

When first entering the bloodstream, white blood cells are filled with dense particles that appear to be lysosomes. If these dense particles *are* lysosomes, how do you predict they will interact with the vesicles containing bacteria?

Prediction 1—Since lysosomes apparently contain the digestive enzymes of the cell, the particles should release their contents into the vesicles containing the bacterium.

Result 1—Observations indicate the dense particles do release their contents into the vesicle. These results suggest the digestive enzymes in the particles (lysosomes) are released at the appropriate location (inside the vesicle) and digest the bacterium.

Question—Do these observations definitely show the dense particles in the white blood cells are lysosomes?

Answer—Although the results provide strong circumstantial support for the hypothesis and for the prediction that isolated particles (lysosomes) correspond to observed cellular particles, there is no *direct* evidence that material in the cellular particles are digestive enzymes. For example, another interpretation could be that digestive enzymes enter the vesicles from some unobserved source other than the dense particles. These dense particles could in fact contain an entirely different material.

Figure 1.1 Electron micrograph of 2 types of particles in the cell. In the middle to lower left are dense particles. The 2 larger, lighter particles to right and center are clearly surrounded by a membrane. *(©Don Fawcett/Photo Researchers, Inc.)*

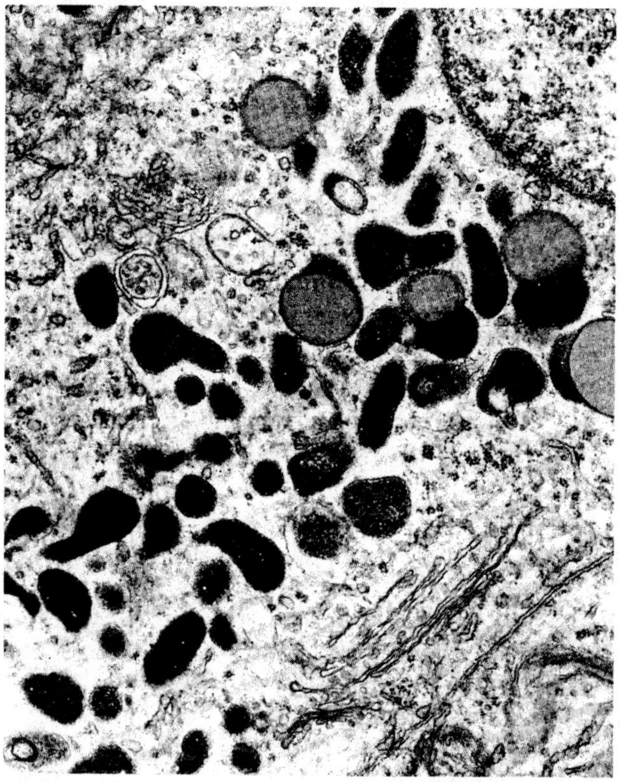

Figure 1.2 Bacterium being engulfed by a white blood cell. *a.* The plasma cell membrane folds around the bacterium. *b.* The bacterium inside the cell and surrounded by a portion of the plasma membrane forming a vesicle. Dense particles fuse with the vesicle.

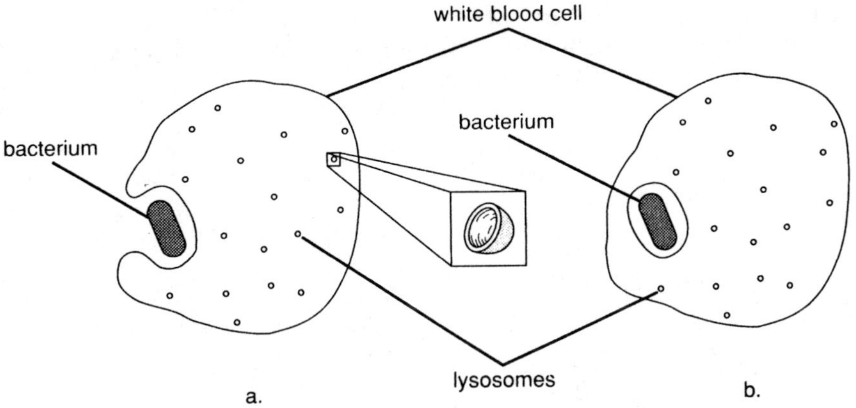

I

In another experiment, material was placed on an electron microscope slide of a cell containing the dense particles. This material turned dark when exposed to digestive enzymes and became visible when viewed with the electron microscope. What would you expect to observe in this experiment?

If isolated lysosomes were mixed with a bacteria culture, what would happen?

Result 1 indicates lysosomes digest foreign material, but do they ever function to digest their host cell? The destruction of animal cells is sometimes necessary as part of the normal developmental process. One of the best known examples is the disappearance of a tadpole's tail as the tadpole develops into a frog. As the tail gradually disappears, the concentration of enzymes from the lysosomes are measured in the cells. What would you expect to find when these concentrations are measured?

Prediction 2—If the enzymes from the lysosomes cause gradual digestion and disappearance of the tail, these enzymes should be found in a high concentration in the interior of the cells.

Result 2—Measurements of enzyme concentrations have shown that the concentration of free lysosomes enzymes gradually increases as the tail becomes smaller. By the time the tail becomes very small, the enzyme concentration is very high even though almost all other cell parts have been completely destroyed. Thus, under conditions where cellular structures are destroyed, digestive enzymes from lysosomes appear to be an important part of the process.

Question—If the tail cells are being destroyed, how can the enzyme concentration be increasing? In other words, how can an increased concentration of enzymes be created?

Answer—The absolute amount of enzymes present does not increase. Because the number of cells is constantly decreasing while the enzymes are not, the concentration of enzymes present in the decreasing number of cells is constantly increasing.

II

What would happen to cells in other parts of the tadpole's body if lysosome enzymes were released in these cells?

If cells in the tail that were in the process of being digested were examined under the electron microscope, would lysosomes be observed?

What would be an advantage of "packaging" the digestive enzymes in lysosomes rather than just producing the enzymes when needed?

Thus, lysosomes appear to be highly specialized organelles that serve to isolate digestive enzymes and release them under appropriate conditions as proposed in the hypothesis. Discovery and the investigation of lysosomes depended on the development of the electron microscope and experimental techniques that aided in revealing the biochemical interactions in the cell.

Other questions remain. For example, how are lysosomes produced? Do all cells contain these organelles? Answering these questions requires further experimentation and analysis.

Reference

de Duve, C. 1965. *The Living Cell: Readings from Scientific American.* San Francisco: W. H. Freeman.

Case Study 1 THE LYSOSOME

Name ______________________
Section ______________________
Date ______________________

I

Question 1—In another experiment, material was placed on an electron microscope slide of a cell containing the dense particles. This material turned dark when exposed to digestive enzymes and became visible when viewed with the electron microscope. What would you expect to observe in this experiment?

Answer 1

Question 2—If isolated lysosomes were mixed with a bacteria culture, what would happen?

Answer 2

II

Question 1—What would happen to cells in other parts of the tadpole's body if lysosomal enzymes were released in these cells?

Answer 1

Question 2—If cells in the tail that were in the process of being digested were examined under the electron microscope, would lysosomes be observed?

Answer 2

Question 3—What would be an advantage of "packaging" the digestive enzymes in lysosomes rather than just producing the enzymes when needed?

Answer 3

2 Case Study: Human Pedigrees (Difficulty Level 1)

Experiments on the inheritance of genetically determined traits most frequently use only a few species of animals and plants, which have characteristics well suited for genetic experiments. For example, fruit flies are often used because they have a short generation time and produce large numbers of offspring. Consequently, a large amount of data can be collected quickly. These organisms can be easily raised in the laboratory and require little space.

Investigating the inheritance of genetic characteristics in humans, however, can be a complicated and difficult study. Several generations in the same family must be observed to obtain a sufficient amount of information to analyze. Another problem is that family size may be small and the inheritance may not follow the expected ratios or percentages if only small numbers of people are available. Moreover, the generation time is relatively long, and if family records are used to gather information on genetic traits in previous generations, information may be missing or incorrect.

In spite of these difficulties, tracing the inheritance of specific traits through several generations has been used extensively to study human inheritance. This ancestral line, or family pedigree, has been kept by many families as a genealogical record so the family ancestry can be traced.

How can family pedigrees help in the study of genetics? Can such pedigrees provide sufficient information to show the characteristics of inheritance in humans? For example, to show whether a trait is dominant or recessive? These questions lead to the following hypothesis:

Hypothesis—Careful analysis of human pedigrees can provide information that will allow conclusions to be drawn about the characteristics of the inheritance.

Testing this hypothesis obviously requires collecting and examining information on a *suspected* genetic trait for several generations. One example of the collection of such information is given in figure 2.1, which indicates the appearance of Huntington disease in 4 generations of a family. Huntington disease is a mental disorder with mental deterioration and uncontrollable muscular movements.

Is this mental disorder an inherited disease? An initial examination of the pedigree chart suggests that it is. The disorder appears in every generation of this family. In the general population, however, the disease is rarely found.

Are there other indications that the disease is inherited? What would you predict would be observed if the trait is dominant?

Prediction 1—If Huntington disease is a dominant trait, it should *not* appear in any offspring unless one or both parents have the disease.

Result 1—Figure 2.1 shows that in every case where a person with Huntington disease married a normal person and produced children, at least one child developed the disease. On the other hand, when neither parent exhibited the disease, no offspring were shown to have the disorder. For example, the couple in the II generation, parents 1 and 2, produced 6 children in the III generation. Of these 6, 5 exhibited the disease and all who married had at least one child with the disease. The other child, who was normal, married and had *no* children with Huntington disease. These observations are a strong indication that the disease is inherited as a dominant trait.

Question—Is it possible that Huntington disease could be a recessive trait?

Answer—Possible, but not very likely because whenever a person with the disease married a person without the disease, some of the children of this marriage had the disease. For this to occur, every person without the disease would have to be a heterozygous carrier of the gene. This would be highly unlikely.

If male 7 in generation III married a female without the disease, could any of their children have the disease?
If Huntington disease is caused by a dominant gene, could a person who does not have the disease be a carrier?
Are males or females more likely to have the disease according to the pedigree chart?

Figure 2.1 Human pedigree of a family showing Huntington disease.

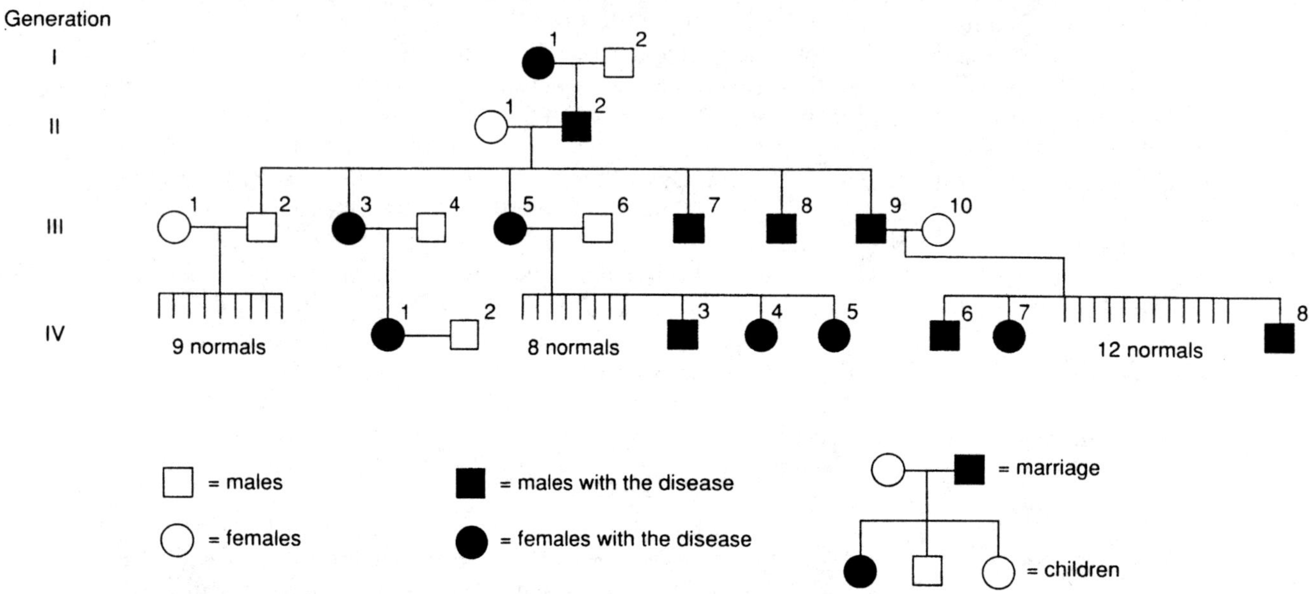

Another way to look at the information in the pedigree is to examine the ratio of people affected by the disease compared to those who are not affected. If Huntington disease is a dominant trait, then what ratio of offspring with and without the disease would you expect when an affected individual marries an individual without the disease?

Any affected individuals will very likely be heterozygous for the disease because they are the offspring of parents, one of whom has the disease and one of whom does not. This cross can be shown as follows:

Hh × *hh* Where *H* is the dominant gene for Huntington disease and *h* is the recessive normal gene.

Any offspring of this cross who carry the dominant gene (*H*) must be *Hh* since only the normal gene (*h*) can be inherited from the normal parent.

In the cross of *Hh* × *hh,* what portion of the offspring should be affected?

Prediction 2—In this cross, the portion of affected offspring can be predicted from the Punnett square:

genes from affected parent \ genes from unaffected parent	*h*	*h*
H	*Hh*	*Hh*
h	*hh*	*hh*

Half the offspring should be affected (*Hh*) and half should be unaffected (*hh*).

Result 2—Examination of the pedigree chart indicates a much greater proportion of people are unaffected than would be expected. There are, however, reasons why this unexpected proportion might occur. First, Huntington disease does not usually develop until a person is at least 40 years old. Thus, a person could have the gene for the disease but if death occurred before the symptoms were apparent, they would not be recorded as a person with Huntington disease. Second, in earlier generations when medical diagnosis was less sophisticated, the disease may have been recorded as a different affliction. Therefore, an above average number of individuals recorded as unaffected would be expected in earlier years.

Question—Why does the pedigree chart not show any marriages between 2 individuals who have the disease?

Answer—Huntington disease is rare outside the families where it is being inherited. Since people normally marry outside their families, the chances of marrying someone who has the disease are remote.

II How many people in the family are recorded as having the disease? (Do not count those who marry into the family.)

How many are recorded as *not* having the disease?

Could a human pedigree be used to detect a recessive trait? How?

Although the pedigree chart does not obviously support prediction 2, there are good reasons for this discrepancy, and the results obtained from an analysis of the chart are accepted as evidence that Huntington disease is a genetic disease inherited as a dominant trait. The analysis also illustrates that available evidence does not always obviously support reasonable predictions. Other factors, such as the late development of the disease, may need to be considered to explain the data. Even with these difficulties, however, pedigree analysis has proven to be extremely valuable for investigating human inheritance, and this example of Huntington disease strongly supports the hypothesis.

Reference

Moody, P. A. 1967. *Genetics of Man.* W. W. Norton.

Case Study 2 HUMAN PEDIGREES

Name ____________________
Section ____________________
Date ____________________

I

Question 1—If male 7 in generation III married a female without the disease, could any of their children have the disease?

Answer 1

Question 2—If Huntington disease is caused by a dominant gene, could a person who does not have the disease be a carrier?

Answer 2

Question 3—Are males or females more likely to have the disease according to the pedigree chart?

Answer 3

II

Question 1—How many people in the family are recorded as having the disease?

Answer 1

Question 2—How many are recorded as *not* having the disease?

Answer 2

Question 3—Could a human pedigree be used to detect a recessive trait? How?

Answer 3

3 Case Study: Cholesterol and Atherosclerosis (Difficulty Level 2)

Atherosclerosis is a disease of the circulatory system in which plaques or thickenings form in the artery wall, narrow the opening in the artery, and restrict the flow of the blood. If a blood clot forms at the plaque site, blood flow can be seriously reduced and may lead to a heart attack. Extensive studies have shown that people with a high blood level of *cholesterol,* a lipidlike molecule insoluble in water, are much more likely to develop atherosclerosis. The plaques formed in the arteries contain large amounts of cholesterol. High levels of cholesterol are found in meat, eggs, and some cheeses. The relation between atherosclerosis and cholesterol has led to a strong recommendation by medical researchers that cholesterol consumption be minimized.

Since the level of blood cholesterol appears to be a very important factor in affecting the development of atherosclerosis, detailed investigations have been made on how blood cholesterol is controlled. High cholesterol in the diet causes an increase in blood cholesterol, but how is it removed from the blood? Studies have shown that particles composed of protein, lipids, and cholesterol (termed *low density lipoprotein* particles, LDL) carry almost all the blood cholesterol. Each particle contains more than 1,500 cholesterol molecules and one protein molecule termed the *LDL protein.*

Cells utilize cholesterol in a variety of ways—as an essential part of membranes, to synthesize some hormones, and in the production of bile salts by the liver that help disperse fats in the digestive system. As the cells utilize cholesterol and convert it to other substances, the blood cholesterol (i.e., the number of LDL particles) decreases. In other words, whatever controls the movement of LDL particles into the body cells controls the removal of cholesterol from the blood.

Since the particles are rather large, how do they get inside the cells? The following hypothesis proposes a mechanism for the uptake of LDL particles:

Hypothesis—The plasma membranes of cells have receptor molecules that can combine with LDL particles. This combination results in endocytosis (the engulfing of relatively large particles by the plasma membrane), and the rate of LDL uptake is controlled by the number of receptor molecules.

If such receptor molecules do exist, how can they be investigated? As mentioned earlier, the LDL particle also contains a protein. Perhaps this protein serves as a recognition site where the receptor molecule on the plasma membrane combines with the LDL particle. If so, what would you predict would happen when the LDL protein is changed in some way?

Prediction 1—If the LDL protein is the recognition site and is changed in such a way that it *cannot* combine with the recognition molecule, LDL particles should be removed from the blood more slowly.

Result 1—LDL particles can be labeled with radioactive molecules and injected into experimental animals. By measuring the amount of radioactivity remaining in the blood at various times after injection, the rate of LDL particle uptake by cells can be determined. Subsequent measurements showed that when the LDL protein was changed, the particles remained in the blood much longer than did particles with unchanged protein.

These results are consistent with the presence of receptor molecules on cell membranes and are therefore consistent with the hypothesis.

Question—What would be a serious criticism of the interpretation of Result 1?

Answer—Result 1 does not directly show the presence of a receptor molecule, only that when the LDL protein is changed, the uptake of LDL particles is decreased. The LDL protein has been shown to be important in uptake, but whether this is because the protein combines with a receptor molecule has not been shown.

Question—Does the decrease of the radioactively labeled LDL particles indicate that plasma membrane receptor molecules are present?
Answer—No. Several mechanisms could account for the decrease of the particles. Plasma membrane receptors are only one such mechanism. For example, the particles could be excreted from the body or they could accumulate on the walls of the arteries and veins without actually leaving the circulatory system.

I If the LDL protein were *completely removed* from the particle, how would uptake be affected? As the radioactivity disappears from the blood, where would it be found?

Other experiments have directly investigated the presence of plasma membrane receptors by isolating the plasma membrane and examining the binding of LDL particles to proteins present in the membrane. When this is done, it is found that binding of LDL particles does occur and the number of receptors varies widely depending on the type of cell.

What types of cells would you predict would have the greatest number of plasma membrane receptors?
Prediction 2—Cells in organs and tissues where cholesterol is most utilized should exhibit the greatest number of receptors.
Result 2—The greatest number of membrane receptors were found in liver cells (where bile salts are produced) and in the ovaries and adrenal gland (where hormones are produced). Thus, as predicted, the greatest number of receptors are found on cells where cholesterol is most utilized.

These results now provide much stronger support for the hypothesis since the presence of membrane receptors has been shown directly and the number of receptors has been shown to be related to cholesterol utilization.
Question—Do you think *all* body cells should have some receptors for LDL particles?
Answer—Cholesterol is an essential component of cell membranes, both in the plasma membrane and in the internal structures, such as the endoplasmic reticulum. All cells then will utilize cholesterol to some extent as part of their membranes and would be expected to have some membrane receptors.

II If radioactive LDL particles are injected into an animal, what organs and tissues would later show the greatest amount of radioactivity?
If a chemical that combined with the receptor molecules on the plasma membrane was injected into an animal, how could this affect cholesterol uptake?

Further investigations have shown that there are people who suffer from a genetic disease (familial hypercholesterolemia) in which the blood cholesterol is constantly at a high level. Even young children with this disease can have heart attacks. Apparently, the disease is caused by a mutation in a single gene that carries information for the membrane receptor protein. About one in 500 people carry the mutation. If the mutation is present, the receptor is defective and cannot bind the LDL particles. As a result, the blood cholesterol remains high because the LDL particles are not removed as effectively from the blood.

The disease appears in 2 forms, a very severe form and a less severe form. Why is it logical that 2 forms of the disease would be expressed? What would you predict would be observed if the genetics were examined of individuals with the very severe form and the less severe form of the disease?
Prediction 3—People with the more severe form should the mutated gene on both homologous chromosomes that carry the gene for the receptor protein. The less severe form should appear in individuals who are heterozygous.

Result 3—Subsequent investigations of individuals suffering from the disease did show that the prediction was accurate.

III What might be done for individuals who have the less severe form of the disease to increase the number of membrane receptors?
Would this approach work for those individuals with the more severe form?
If radioactive LDL particles were injected into normal, heterozygous, and homozygous individuals, how quickly would the radioactivity decrease in the circulatory systems of each of these individuals?

These studies of the genetics of familial hypercholesterolemia have produced further support for the hypothesis. Thus, investigations of the cellular mechanisms of cholesterol uptake, receptors on the plasma membrane, and genetic experiments have all provided evidence supporting control of blood cholesterol by receptor mediated uptake.

Reference

Brown, M. S., and Goldstein, J. L. November 1984. How LDL receptors influence cholesterol and atherosclerosis. *Scientific American,* 58-66.

Case Study 3 CHOLESTEROL AND ATHEROSCLEROSIS

Name ____________________
Section ____________________
Date ____________________

I

Question 1—If the LDL protein were *completely removed* from the particle, how would uptake be affected?

Answer 1

Question 2—As the radioactivity disappears from the blood, where would it be found?

Answer 2

II

Question 1—If radioactive LDL particles are injected into an animal, what organs and tissues would later show the greatest amount of radioactivity?

Answer 1

Question 2—If a chemical that combined with the receptor molecules on the plasma membrane was injected into an animal, how could this affect cholesterol uptake?

Answer 2

III

Question 1—What might be done for individuals who have the less severe form of the disease to increase the number of membrane receptors?

Answer 1

Question 2—Would this approach work for those individuals with the more severe form?

Answer 2

Question 3—If radioactive LDL particles were injected into normal, heterozygous, and homozygous individuals, how quickly would the radioactivity decrease in the circulatory systems of each of these individuals?

Answer 3

4 Case Study: Chromosome Structure (Difficulty Level 2)

Chromosomes are puzzling structures in the cell. Chromosome structure and its changes have long intrigued scientists. For most of the life of the cell, the chromosomal material is not visible with the compound light microscope. Only when the cell is nearing cell division and during cell division do the chromosomes "condense" to visible structures. When cell division is completed, the chromosomes in the daughter cells again disperse and are not visible. Initial attempts to understand this structure were conducted by studying its molecular constituents.

Chromosomes are composed of *chromatin,* which is comprised of DNA and protein. Although all the genetic information is carried by DNA, more than 50% of the chromatin is protein. What is the function of this protein? What is the relationship between the DNA and protein? Can investigations of this relationship provide information on chromosome structure?

Investigations of the protein component have indicated that it is comprised of histone molecules (a negatively charged structural protein). These same histones are found as part of the chromatin in almost *all* eukaryotic organisms. Such a universal presence suggests that histones play a critical role in chromosome structure and function.

Early experiments on the structure of chromatin used *X-ray diffraction,* a technique in which X rays are beamed at a sample of the material being investigated. Scattering of the X rays can provide a measure of any repeating structure in the material. With chromatin, a repeating structure about every 100 angstroms was indicated. This was a somewhat surprising result because the DNA is a long, thin, double helix with no repeating structures.

How does the DNA-histone combination in chromatin result in a repeating structure? Such initial questions and the results led investigators to propose the following hypothesis:

Hypothesis—The DNA-histone combination of chromatin forms an organized structure that can account for the condensation of the chromosome.

Testing this hypothesis involved a variety of experiments. Some of the most informative involved treating chromatin and DNA with enzymes capable of cleaving the DNA molecule. By using several different enzymes with different properties, various aspects of the DNA-histone structure could be investigated.

In one experiment, DNA and chromatin were treated with a DNA-cleaving enzyme extracted from rat liver cells. Since chromatin is a combination of DNA and histone, it is possible the histone could protect some part of the DNA from cleavage.

What would you predict would be observed when purified DNA and chromatin were exposed to the enzyme?

Prediction 1—If histone protects the DNA from cleavage, DNA in chromatin should be cleaved into fragments of different lengths when compared to the fragment lengths from treatment of the purified DNA.

Result 1—Cleavage of purified DNA resulted in random fragment lengths. Cleavage of chromatin, however, resulted in DNA fragments that were multiples of a smallest unit length of about 200 nucleotides. In other words, the only fragments found from chromatin were in lengths of one unit length, 2 unit lengths, 3 unit lengths, and so on. These results are consistent with the prediction that histones protect the DNA from cleavage. The observation that the DNA fragments are produced in unit lengths suggests that the histone protecting the DNA is arranged in some regular or periodic way on the DNA, much like histone beads on a DNA thread. The DNA between the histone beads would be exposed and could be cleaved, while the histone beads would protect short fragments (i.e., the unit length of DNA) from cleavage. Thus, these results also provide a potential explanation of the X-ray diffraction experiments. A regular histone arrangement could provide a repeating structure as indicated by the X-ray studies.

Question—Why was the DNA in histone cleaved into *multiples* of unit length rather than to all fragments of a single unit length?
Answer—It is possible that if the chromatin were treated with the enzyme for a sufficiently long time, *all* fragments of a single unit length would have been produced. Within a relatively short time exposure, however, many DNA sites that could be cleaved may escape contact with the enzyme and result in fragments of a multiple unit length.

I Why should exposure of purified DNA to the enzyme result in DNA fragments of varying, random lengths?
If chromatin were exposed to the enzyme for only a very short time, how would this affect the length of DNA fragments?

A concern that could be raised about the interpretation of Result 1 is that perhaps the DNA unit lengths observed were a result of some property of the particular enzyme used. Would similar results be observed with a DNA cleaving enzyme from another source? In another experiment, chromatin cleavage was studied with an enzyme extracted from a bacteria. Moreover, the exposure to the enzyme was maintained for a much longer time period than in the first experiment.

What results would you predict from this experiment?
Prediction 2—If the unit length DNA fragments are a result of the chromatin properties, results similar to Result 1 should be observed.
Result 2—Extensive treatment of chromatin with the bacterial DNA cleaving enzyme resulted in DNA fragments of approximately the same unit length as those observed in Result 1. Moreover, essentially all of the DNA was reduced to the unit length fragment size of 200 nucleotides. A further puzzling result was that the basic DNA-histone unit was about 100 angstroms in diameter as shown by X-ray diffraction, but the 200 nucleotide fragment was some 700 angstroms in length. How do the DNA and histone interact to contain a 200 nucleotide length in the 100 angstrom structure?

Both Result 1 and Result 2 support the hypothesis by indicating that the DNA-histone combination forms an organized structure, but they provide little information on how the chromatin might be folded into a condensed chromosome structure.

II If Result 1 *had* been due to specific properties of the enzyme extracted from rat liver, what would be the expected observations in Result 2?

A third experiment utilizing a DNA-cleaving enzyme from the pancreas was used to investigate further the DNA-histone structure. An important property of this enzyme is that it can cleave the DNA molecule when the DNA is against a solid material such as the histone. The other enzymes, however, can easily cleave the DNA only when it is *not* against a solid material.
Prediction 3—If the enzyme can easily cleave the DNA when the DNA is against a solid support, and if the DNA is on the *outside* of the histone, short DNA fragments should be produced. If the histone *surrounds* the DNA, however, only 200 nucleotide unit DNA fragments should be found.
Result 3—Treatment with the enzyme from the pancreas resulted in single-stranded DNA fragments about 10 nucleotide units in length. These short DNA fragments would be expected if the DNA molecule was on the outside of the histone. This result provides additional information consistent with the hypothesis.

III **How does the finding that DNA is around the outside of the histone help to explain how 700 angstroms of the DNA molecule can be contained in a 100 angstrom histone subunit?**
Why was it necessary to use *3 different* kinds of DNA cleaving enzymes to reveal the information in Results 1, 2, and 3?
If the histone molecule structure were further investigated, what questions might be asked?

Figure 4.1 Model of chromatin and chromosome structure derived from experimental studies.

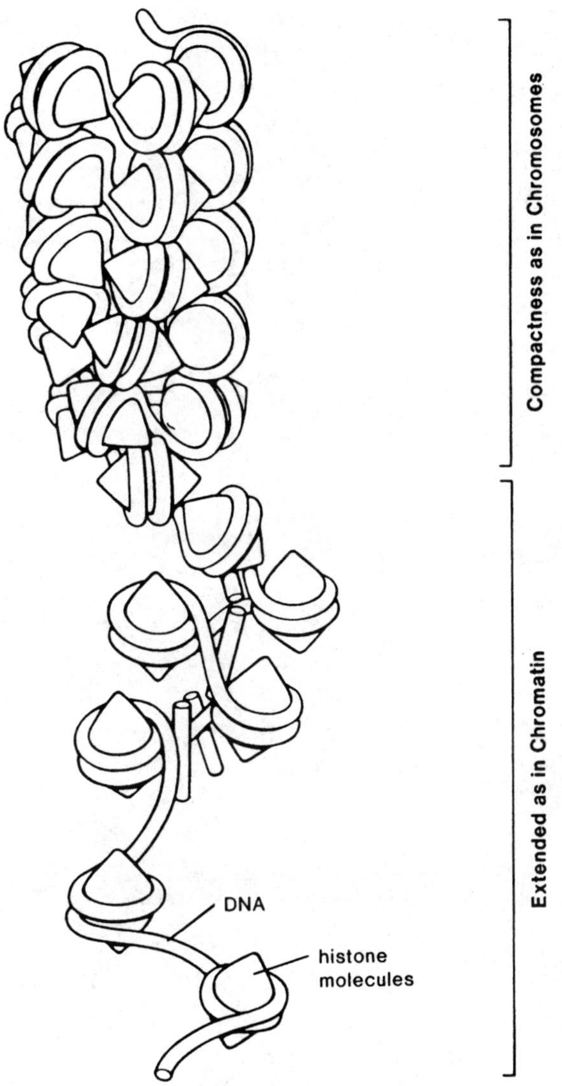

These sequential experiments have added increasing detail and continue to support the hypothesis of the model of chromosome structure. The model that emerges has incorporated these as well as additional results from studies of histone structure and X-ray diffraction (fig. 4.1). Close packing of the DNA-histone units can explain the condensation of the chromatin to form the visible chromosome structure. To arrive at this relatively simple model has required a variety of investigations with considerable analysis of experimental data.

Reference

Kornberg, R. D., and Klug, A. February 1981. The nucleosome. *Scientific American,* 52–79.

Case Study 4 CHROMOSOME STRUCTURE

Name ____________________
Section ____________________
Date ____________________

I

Question 1—Why should exposure of purified DNA to the enzyme result in DNA fragments of varying, random lengths?

Answer 1

Question 2—If chromatin were exposed to the enzyme for only a very short time, how would this affect the length of DNA fragments?

Answer 2

II

Question 1—If Result 1 *had* been due to specific properties of the enzyme extracted from rat liver, what would be the expected observations in Result 2?

Answer 1

III

Question 1—How does the finding that DNA is around the outside of the histone help explain how 700 angstroms of the DNA molecule can be contained in a 100 angstrom histone subunit?

Answer 1

Question 2—Why was it necessary to use *3 different* kinds of DNA cleaving enzymes to reveal the information in Results 1, 2, and 3?

Answer 2

Question 3—If the histone molecule structure were further investigated, what questions might be asked?

Answer 3

5 Case Study: Down Syndrome (Difficulty Level 2)

Individuals with Down syndrome typically exhibit mental retardation and carry a high risk of developing cataracts and leukemia, as well as a variety of other abnormalities. Although Down syndrome has long been recognized as a disability characterized by mental retardation, it was only in the 1950s that chromosomal abnormalities were connected to the disease. Detailed observations indicated in every case that the cells of affected individuals contained 3 number 21 chromosomes.

As shown in figure 5.1, human cells normally contain 23 homologous pairs of chromosomes for a total of 46. The 2 members of each pair are identical, and each numbered pair can be identified by size and shape. Notice that only in the number 21 group are 3 chromosomes present.

Figure 5.1 Karyotype of the chromosomes from an individual with Down syndrome. Chromosomes occur in pairs in normal individuals. Paired chromosomes are also observed in this individual with the exception of group number 21, where 3 chromosomes are present. Apparently, the presence of this one extra chromosome leads to the variety of abnormalities in Down syndrome. *(©Biophoto Associates/Photo Researchers, Inc.)*

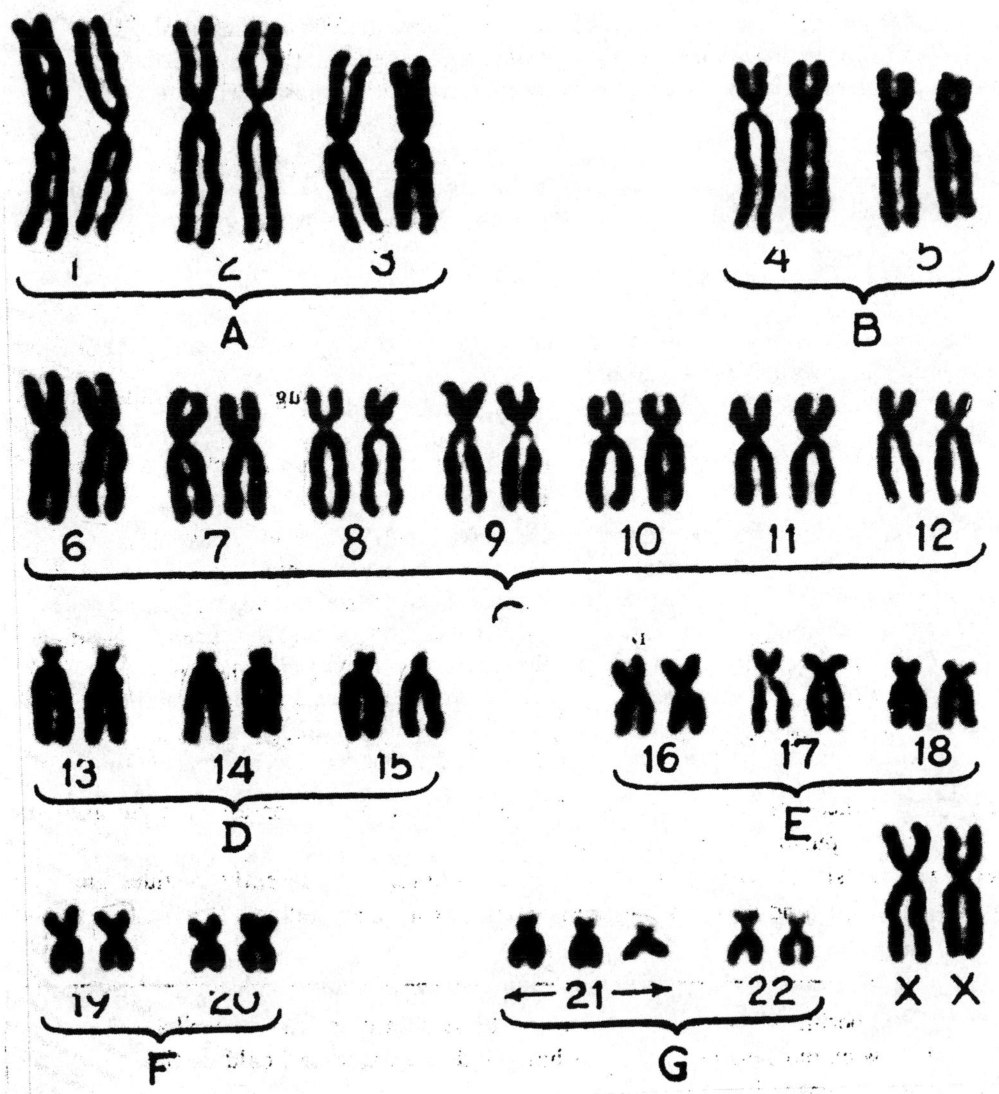

In some cases, Down syndrome is caused by only one extra part of a chromosome 21. In about 5% of the instances, part of a chromosome 21 becomes translocated and attached to another chromosome. In this way, each cell of the affected individual contains 2 whole chromosomes and at least part of another chromosome 21. In every instance of Down syndrome, however, affected individuals show extra chromosome 21 material, in whole or in part.

How does this extra chromosomal material result in the variety of effects characterizing Down syndrome? Possessing extra chromosomal material would mean that extra copies would be present of those genes found on chromosome 21 (or parts of these chromosomes). A logical hypothesis is as follows:

Hypothesis—Genes found on chromosome 21 lead to the variety of specific characteristics of Down syndrome.

Testing this hypothesis, however, requires identifying the genes present and the effect of the gene products on the human body. Although chromosome 21 is relatively small (see fig. 5.1), as many as 1,500 genes are estimated to be present, but less than 20 of these have been identified.

One method that has been used to identify genes present on chromosome 21 is to investigate the amount of some gene product present in an individual with Down syndrome. How would you predict the amount would be changed relative to a normal individual?

Prediction 1—Since affected individuals have 3 rather than the usual 2 copies of a gene, the amount of gene product produced should be higher than in normal individuals.

Result 1—Increased concentrations of free purines, 2 of the bases found in nucleic acids, are observed in persons with Down syndrome. Elevated purine levels can cause mental retardation. Thus, a higher level of a gene product (purines) can explain the occurrence of one characteristic of Down syndrome.

Question—What assumption was made in Prediction 1?

Answer—The prediction assumed all 3 copies of the gene are being expressed. The results suggest the assumption was valid, but there was no way of knowing this would be the case when the prediction was made.

Is it possible that some genes on chromosome 21 *will not* be detected by an increased level of gene product in individuals with Down syndrome?

If *4* copies of chromosome 21 are present in an individual, what would you find when the purine level is measured?

Are the genes directing purine synthesis present on chromosome 21? Result 1 clearly suggests that this should be the case, but how could stronger evidence be obtained?

Investigations of these questions have utilized "fused" cells. With special treatment, human cells and Chinese hamster cells can be induced to fuse, forming a *single* cell containing the chromosomes from both cells. These cells can then be cultured on a artificial growth medium. For reasons not clearly understood, the fused cells often lose most of the human chromosomes and will sometimes result in a cell containing only a single human chromosome 21 in addition to the hamster chromosomes. The particular strain of hamster cells used in these experiments has a mutation that prevents the original hamster cells from synthesizing purines. These cells will *not* grow and reproduce unless grown on a medium containing purines. Fused cells that contain both human and hamster chromosomes, however, can synthesize purines and will grow on artificial medium that does *not* contain purine.

How can these experiments be used to establish that the genes responsible for the synthesis of purines are present on chromosome 21?

Prediction 2—If the gene(s) for purine synthesis are on chromosome 21, the fused cells that contain *only* chromosome 21 from humans should be able to grow on a purine-free growth medium.

Result 2—Subsequent experiments were consistent with the prediction. Chromosome 21 apparently provides the genetic information for the synthesis of purines, and the cells can grow on the purine-free medium.

II If the fused cells used in the experiment contained a few human chromosomes as well as chromosome 21 and could grow on purine-free medium, what could you conclude would be the location of the genes coding for purine synthesis?
Could a fused cell that contained about half the human chromosomes *fail* to grow on purine-free medium? How?

Can other physical and biochemical characteristics of Down syndrome be related to genes on chromosome 21? Since the genes coding for purine synthesis have been located on chromosome 21, others related to Down syndrome should also be found. If so, this finding would provide further support for the hypothesis.
Prediction 3—Additional genes will be found on chromosome 21 that carry the genetic information for products that can be related to Down syndrome.
Result 3—The gene for a protein (alpha-A-crystallin protein) that is a component of the lens of the eye has been found on chromosome 21. This result could explain why individuals with Down syndrome are more likely to have cataracts and lens defects. A cancer-causing gene (oncogene) has also been discovered on chromosome 21 and could be related to the increased risk of leukemia. Most recently, the gene coding for one of the major proteins accumulating in the brain of individuals with Alzheimer's disease has been located on chromosome 21. This protein is also found in high concentration in the brain of those affected with Down syndrome.

III From the information presented in Result 1, what would you expect to find when the concentration of alpha-A-crystallin protein is measured in a person with Down syndrome?
Could experiments with fused cells be used to confirm the presence of the gene for alpha-A-crystallin protein on chromosome 21? How?

Evidence continues to accumulate in support of the hypothesis that the genes on chromosome 21 are directly related to the characteristics of Down syndrome. As indicated earlier, however, fewer than 20 of the 1,500 genes on chromosome 21 have been identified. Continued investigation of these genes can potentially reveal even greater information on the relation between genes and genetic disease expression.

Reference

Patterson, D. August 1987. The causes of Down syndrome. *Scientific American,* 52–61.

Case Study 5 DOWN SYNDROME

Name ____________________
Section ____________________
Date ____________________

I

Question 1—Is it possible that some genes on chromosome 21 *will not* be detected by an increased level of gene product in individuals with Down syndrome?

Answer 1

Question 2—If *4* copies of chromosome 21 are present in an individual, what would you find when the purine level is measured?

Answer 2

II

Question 1—If the fused cells in the experiment contained a few human chromosomes as well as chromosome 21 and could grow on purine-free medium, what could you conclude about the location of the genes coding for purine synthesis?

Answer 1

Question 2—Could a fused cell that contained about half the human chromosomes *fail* to grow on purine-free medium? How?

Answer 2

III

Question 1—From the information presented in Result 1, what would you expect to find when the concentration of alpha-A-crystallin protein is measured in a person with Down syndrome?

Answer 1

Question 2—Could experiments with fused cells be used to confirm the presence of the gene for alpha-A-crystallin protein on chromosome 21? How?

Answer 2

6 Case Study: Oncogenes and Cancer (Difficulty Level 2)

One of the most striking discoveries in the investigation of cancer has been the finding that some viral infections can transform normal cells to cancerous cells. How can a virus carry out this transformation?

When a virus infects a cell, the nucleic acid of the virus uses the cell's biosynthetic mechanisms to synthesize new viral organisms. Relatively few genes are carried on the small amount of viral nucleic acid, and the protein coat of the virus does not appear to have any significant effect on the infected cell. How can these simple components cause transformation to a cancerous state?

Since the viral genes would be the most active component, the following hypothesis has been proposed.

Hypothesis—Specific genes have the ability to convert normal cells to cancerous cells. Such genes have been termed *oncogenes.* Testing this hypothesis has led to very surprising results and to new directions in cancer research.

Interesting initial observations were made on viruses that are able to transform cells. First, all cancer-causing viruses are retroviruses carrying a core of RNA as the genetic material rather than DNA as in other viruses. Retroviruses also carry an enzyme (reverse transcriptase) that can transcribe RNA to DNA. This DNA then directs the synthesis of new viral components.

Second, the DNA transcribed from the RNA is, in most cases, incorporated into the infected cell's DNA. If the infected cell is not destroyed by the virus, this DNA will replicate when the infected cell's chromosomes are duplicated in cell division. Thus, copies of the viral DNA can be passed along to all daughter cells and are a permanent part of the cell's genetic information.

While these characteristics are interesting, they provide no evidence that oncogenes exist. How could the hypothesis be tested?

It is logical first to examine the genetic material of the cancer-causing virus. Such an examination can be carried out using deletion mutants of a cancer-causing virus that have lost the ability to transform normal cells to cancerous cells.

If a deletion mutant is compared to a normal virus, what would you expect to find?

Prediction 1—If an oncogene is responsible for the virus' ability to cause cancer, then the deletion mutant should have an altered or missing gene when compared to the normal virus.

Result 1—A comparison of the genetic material from the deletion mutant and the normal virus revealed the deletion mutant was missing a segment of the viral RNA. This result suggests the missing segment is the oncogene.

Question—What would be a better experiment to test for the existence of oncogenes?

Answer—Result 1 is indirect evidence of the presence of an oncogene and shows only that a difference in the genetic material is somehow related to a difference in the ability to transform cells. Much more direct evidence for oncogenes could be obtained if the RNA segment from a normal virus (which is missing in the deletion mutant) could be shown to transform cells.

Note: This experiment has been done using DNA transcribed from the RNA segment. When injected into normal cells, this DNA fragment causes transformation to the cancerous state.

I If the missing RNA segment were added back to the deletion mutant, how would this affect the ability of the virus to transform normal cells into cancerous cells?

Does the oncogene from the virus become incorporated in the host cell's DNA? If so, is the oncogene still capable of transforming the cell?

Investigating these questions required devising a way to locate the incorporated gene in the DNA of the host cell. An effective method was developed using the following procedure:

1. The RNA from a normal cancer-causing virus was extracted and transcribed to DNA using reverse transcriptase. Radioactive DNA nucleotides were used in the transcription process so the DNA that was synthesized was radioactively labeled.
2. The radioactive DNA was separated into single strands by heating, and the DNA molecule was cleaved into relatively short fragments using a cleaving enzyme.
3. The radioactive DNA fragments were combined with extracted RNA from deletion mutants of the virus. DNA fragments will *hybridize,* or combine, by forming hydrogen bonds with complimentary sequences of base pairs on the RNA. The DNA fragment containing the oncogene, however, will *not* hybridize. Since this segment is not present in the deletion-mutant, there will be no complementary sequence of base pairs where the oncogene DNA fragment can hybridize.
4. The unhybridized DNA fragments were then isolated. What do you predict would be observed when the DNA oncogene-containing fragments were added to single-stranded host cell DNA that had previously been infected by the virus?

Prediction 2—If DNA transcribed from viral RNA is incorporated into host cell DNA, the radioactive DNA should hybridize with the complementary site (containing the incorporated oncogene) on the host cell DNA.

Result 2—As predicted, the radioactive DNA did combine at specific sites on the host cell DNA.

An additional surprising result, however, was that the DNA fragment would hybridize with sites on DNA from *previously uninfected cells!* These first experiments were performed with birds, but later investigations found existing oncogenes in the DNA of mammals and fish, demonstrating that oncogenes are widespread in animals. Apparently, these cells typically contain oncogenes very similar or identical to the viral oncogenes.

Question—What hypotheses could explain how the viral oncogene came to be present in previously uninfected cells?

Answer—One hypothesis would be that the DNA is carried in the animals' DNA from viral infections at an earlier stage in the life of the organism or from infections in previous generations. If the virus inserted DNA into the animal cells' DNA, and the cells survived the infection, the genetic information would be passed on to all daughter cells as these cells reproduced.

A second hypothesis is that the oncogenes are a natural part of the host cells' DNA and have been picked up by a virus in the process of reproduction in the cells.

Question—What are the strengths and weaknesses of the 2 preceding hypotheses?

Answer—Both provide logical mechanisms to explain the presence of the oncogene in the host cell DNA based on the reproductive cycle of the virus and of DNA replication.

A disadvantage of the first hypothesis, however, is that viral infections are probably not so widespread that oncogenes would be inserted in a great many body cells. To be passed along to another generation, the oncogene must be inserted in the sperm or egg cells of a parent. While these occurrences are all possible, they may be unlikely.

The second hypothesis raises puzzling questions. Why should animal cells carry a gene that can cause cancer? Do oncogenes specify other necessary functions in addition to causing cancer? These questions have no obvious answer.

II What if the radioactive DNA synthesized from genetic information in the normal virus were combined with RNA from *another* normal virus rather than RNA from the deletion mutant? Why were the DNA fragments separated into single strands before mixing with RNA in step 2?

If apparent oncogenes normally exist in animal cells' DNA, can these genes cause cancer? This question was investigated in an experiment using mouse cells growing in tissue culture. Normally, the cells—designated 3T3 mouse cells—grow in a single layer on the culture medium and cannot cause tumor growth when injected into mice. DNA from human tumors (which were *not* caused by viral infection) are injected into these mouse cells.

After this treatment, what would you expect to happen to the mouse cells?

Prediction 3—If oncogenes are present in the human tumor cells, the injected human cell DNA should transform the mouse cells from normal to cancerous cells.

Result 3—After DNA injection, the 3T3 mouse cell growth became chaotic on the culture medium and could cause tumors when injected into mice. The tumor cell DNA could cause these cells to become cancerous.

III What if the human tumor in the experiment had been caused by viral infection? How would this change the interpretation of the experiment?

It has been suggested that even though oncogenes may be carried by human cells, these genes must be mutated or activated to cause cancer. Why is this a logical suggestion?

Why were human *tumor* cells used to obtain DNA to inject into 3T3 mouse cells rather than DNA from normal cells?

Results from studies on the ability of viruses to cause cancer have provided considerable support for the hypothesis that specific genes have the ability to convert normal cells into cancerous cells. Moreover, several extremely interesting questions have been raised. Why should the DNA of animal cells, including human cells, contain oncogenes? Do these genes serve some essential functions separate from their cancer-causing properties? Can the cancer-causing ability of oncogenes be altered to provide an important means of controlling cancer? These questions are under intensive investigation and may lead to new treatments of cancer.

References

Bishop, J. M. March 1982. Oncogenes. *Scientific American,* 80–93.

D'Eustachio, P. January-February 1984. Gene mapping and oncogenes. *American Scientist,* 32–40.

Case Study 6 ONCOGENES AND CANCER

Name ____________________
Section ____________________
Date ____________________

I

Question 1—If the missing RNA segment were added back to the deletion mutant, how would this affect the ability of the virus to transform normal cells into cancerous cells?

Answer 1

II

Question 1—What if the radioactive DNA synthesized from genetic information in the normal virus were combined with RNA from *another* normal virus rather than RNA from the deletion mutant?

Answer 1

Question 2—Why were the DNA fragments separated into single strands before mixing with RNA in step 2?

Answer 2

III

Question 1—What if the human tumor in the experiment had been caused by viral infection? How would this change the interpretation of the experiment?

Answer 1

Question 2—It has been suggested that even though oncogenes may be carried by human cells, these genes must be mutated or activated to cause cancer. Why is this a logical suggestion?

Answer 2

Question 3—Why were human *tumor* cells used to obtain DNA to inject into 3T3 mouse cells rather than DNA from normal cells?

Answer 3

7 Case Study: Transport Across Membranes (Difficulty Level 2)

Active transport is the movement of material against a concentration gradient and requires cellular energy (ATP) to carry out this movement.

Sodium ion (Na^+) is one substance maintained at a low concentration inside cells even though fluid bathing the cell has a high Na^+ concentration. As shown in figure 7.1, even though Na^+ concentration is high outside the cell, active transport maintains a high Na^+ efflux and thus keeps Na^+ low inside the cell.

Figure 7.1 Even though Na^+ concentration is high outside the cell, active transport maintains a high Na^+ efflux and thus keeps Na^+ low inside the cell.

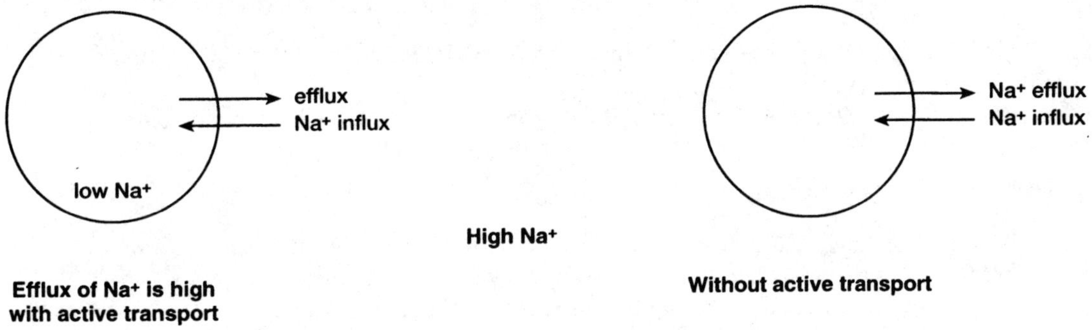

The data shown in figure 7.2 have been proposed as strong evidence supporting the following hypothesis:

Hypothesis—Na^+ is actively transported out of cells.

There are several aspects to these data, and each can be considered in relation to the hypothesis.

Figure 7.2 The data shown have been proposed as strong evidence supporting the hypothesis that Na^+ is actively transported out of cells. There are several aspects to these data, and each can be considered in relation to the hypothesis.

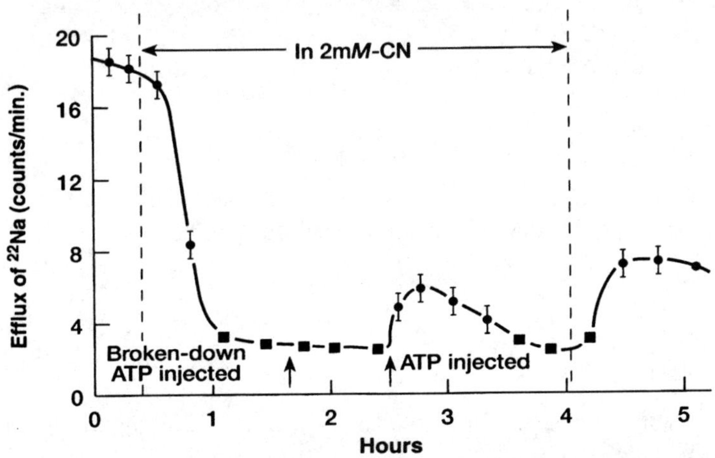

The efflux (movement out of the cell) of Na^+ is measured by injecting radioactive Na^+ into the cell and then measuring how rapidly the radioactivity appears in an external bathing solution. When the cell's metabolism is poisoned, what would you expect to happen to Na^+ efflux?
Prediction 1—If Na^+ is being actively transported out of the cell, then the metabolic poison should stop the production of ATP and thus decrease the Na^+ efflux.
Result 1—Figure 7.2 shows that when the metabolic poison cyanide (CN) is added, the Na^+ efflux drops to about one-fifth of the previous (unpoisoned) rate, and this result is thus consistent with the hypothesis.
Question—What if sodium influx were measured? Would you expect to see the same result?
Answer—The efflux is the movement of Na^+ from a low to high concentration, and net movement would require energy. On the other hand, the influx would be expected to be passive, with no energy requirement, and CN should have no effect.

Adding the CN has presumably stopped the cells' ability to produce ATP and thus eliminated the energy source for active transport. What do you predict would happen if ATP were added back to the cell?
Prediction 2—If Na^+ is being actively transported, then adding ATP should "restart" the active transport abolished by adding CN.
Result 2—As shown in figure 7.2, the ATP added at 2 1/2 hours increases the efflux of Na^+ and again supports the hypothesis.

I When ATP was added at 2 1/2 hours, the Na^+ efflux did not increase to the original level but increased only temporarily and then decreased. Why do you think this occurred?
What could have been done to increase the Na^+ efflux beyond the level observed at 2 1/2 hours?

In another treatment, ATP that had been boiled to break the terminal, high-energy bond was applied to the cells. When this is done, what would be the obvious prediction and why?
Prediction 3—Since the terminal phosphate bond has been broken by boiling, the boiled ATP would not provide energy for active transport, and no increase in efflux should be observed.
Result 3—When boiled ATP is added at 1 1/2 hours, no efflux increase is observed. This result also indicates that it is the energy provided by the ATP molecule that is the key contribution to the active transport process and not some other component of the ATP molecule.

II Suppose some component of the ATP molecule other than its energy contribution *was* affecting active transport. What would have been observed in Result 3?
Result 3 also shows that the physical process of applying some material to the cells does not affect efflux. How does it show this?

The results shown in figure 7.2 represent one of the most conclusive experiments on active transport. This experiment indicates how much can be revealed by a single, well-designed and carefully conducted experiment.

References

Keynes, R. D. 1961. "The Energy Source for Active Transport in Nerve and Muscle." In *Membrane Transport and Metabolism,* 131–39. New York: Academic Press.
Stein, W. D. 1967. *The Movement of Molecules Across Membranes.* New York: Academic Press.

Case Study 7 TRANSPORT ACROSS MEMBRANES

Name ____________________
Section ____________________
Date ____________________

I

Question 1—When ATP was added at 2 1/2 hours, the Na^+ efflux did not increase to the original level but increased only temporarily and then decreased. Why do you think this occurred?

Answer 1

Question 2—What could have been done to increase the Na^+ efflux beyond the level observed at 2 1/2 hours?

Answer 2

II

Question 1—Suppose some component of the ATP molecule other than its energy contribution was affecting active transport. What would have been observed in Result 3?

Answer 1

Question 2—Result 3 also shows that the physical process of applying some material to the cells does not affect efflux. How does it show this?

Answer 2

8 Case Study: Metastasis of Cancer Cells (Difficulty Level 3)

Cancerous tissue is comprised of cells that exhibit irregular, uncontrolled growth. As a result of this growing mass of abnormal tissue, normal body processes are disrupted and death of the organism is probable. Apparently, cancerous cells develop from normal cells due to viral infection or mutation. If the mass of cancerous tissue or tumor remained localized, effective treatment could be applied simply by surgical removal. A primary difficulty, however, is that cancerous cells may move to other parts of the body via the blood vessels and lymph system and set up secondary tumors. Such movement is termed *metastasis.* These secondary growths usually become so widespread that treatment by removal is impossible.

Numerous investigators have examined metastasis with the goal of better understanding the phenomenon and discovering ways of prevention. In a revealing study by Fidler and Kripke in 1970, a skin cancer that has a tendency to metastasize to the lungs was investigated. Cells were removed from a tumor in the lung of a mouse. These cells were separated, and each cell was cultured on artificial media to form separate colonies or clones. Cells from each clone were injected into separate mice, and after a period of growth, the number of resulting lung tumors was counted.

The number of tumors generated by different clones was highly variable, ranging from no tumors produced in some clones to as many as 500 with others. Clearly, the clones (representing individual cells from the original lung tumor) varied widely in their ability to metastasize.

In a subsequent study, cells from a single clone were separated, cultured separately in subclones, and injected into mice.

What would you expect to observe in this case?

Prediction 1—Since the cells of a clone grew from a single original cell, similar results should be observed in the ability to metastasize.

Result 1—

Source	Number of Lung Tumors (median from several injections)
Parent clone	18
Subclone 1	6
2	24
3	39
4	42

(Data from Fidler and Kripke, 1970)

The number of tumors formed by subclones was not significantly different from the parent clone. Similar data were obtained from several other parent clones. The results indicate that the variability in ability to metastasize was due to cellular variation in the parent tumor, not to variability of cells in a clone.

I What possible alternative interpretation of the results from the original experiment does Result 1 eliminate?

Based on these results indicating variability, it was hypothesized that cells that tend to metastasize can be differentiated from cells that do not tend to metastasize by biochemical and/or metabolic differences. Moreover, since a metastasizing cell must move through the blood and lymph system while avoiding capture and destruction by the body's immune system, characteristics that determine immunological interactions were obvious choices for examination.

Recognition of cells by the immune system is partially controlled by molecules found on the cell surface that are part of the major histocompatibility complex (MHC). These molecules are different in every individual and enable the immune system to identify a cell as foreign or as belonging to the individual. Cancer cells, however, undergo a number of changes that can result in their being recognized as foreign even though they develop from the body's own cells.

One series of experiments investigated 2 cell-surface MHC molecules, designated as H-2K and H-2D, which occur in almost all cells. To test whether H-2K and H-2D might be involved in the ability of tumor cells to metastasize, the amount of these MHC molecules was measured in 2 clones. The cells of one clone normally exhibited a high degree of metastasis while another exhibited almost no metastasis.

What would you predict as a result of this investigation?

Prediction 2—If H-2K and H-2D molecules are involved in the cell's ability to metastasize, a difference between metastatic and nonmetastatic cells should be found in these molecules.

Result 2—The nonmetastatic cells exhibited approximately equal amounts of H-2K and H-2D, while the cells showing higher metastasis exhibited almost no H-2K but a large amount of H-2D. These differences are clearly consistent with the hypothesis and suggest further experiments.

II What if *no* difference in the H-2D and H-2K molecules was found in metastasizing and nonmetastasizing cells? Would this indicate that H-2D and H-2K are not involved in metastasis? Why or why not?

Just because there is an observed difference in H-2D and H-2K, however, does not indicate that these molecules invoke a different response in the body's immune response to cancer cells. To investigate this matter further, cells from a highly metastatic clone and a nonmetastatic clone were injected into mice and the immune response was observed. These cloned tumor cells were subjected to X rays before injection so they would not form tumors, and the mice used for injection were a specially bred variety that had MHC molecules identical to those of the tumor cells. (This was done to insure that the mice's immune system did not respond to different MHC molecules rather than to the H-2K and H-2D proportions.) After 2 weeks, lymphocytes were extracted from the mice and exposed to tumor cells grown in a culture.

Prediction 3—An immune response will be observed with the lymphocytes extracted from mice injected with nonmetastasizing cells but not with lymphocytes extracted from the other mice.

Result 3—In subsequent experiments the predicted results were observed. Apparently, cells that exhibit low metastasis do so because the body's immune system can recognize and destroy these cells. Tumor cells that are highly metastatic exhibit this property because they escape detection and destruction by the immune system.

III Why was it important that the tumor cells *not* form tumors when injected into the mice? Why couldn't this experiment have been performed by observing lymphocytes in the injected mice rather than the more complicated procedure of extracting lymphocytes and checking for interaction with cultured tumor cells?

In a subsequent experiment, cells that had a low tendency to metastasize were exposed to a chemical that caused the expression of several genes, including the one coding for the H-2D molecules. Thus, cultured tumor cells that normally exhibited a relatively low concentration of H-2D were stimulated to express a relatively high concentration of H-2D.

What would you predict would happen if cells altered in this way were injected into mice?

Prediction 4—If a higher concentration of H-2D decreases recognition of the cancer cells by the body's immune system, then injection should result in a larger number of tumors formed by these altered cells.

Result 4—Results from this experiment indicated that, as predicted, injection of the altered cells resulted in a significantly larger number of tumors compared to injections of untreated tumor cells.

IV Why was this experiment done? What additional information did it provide beyond Result 3? What is a primary deficiency of this experiment (i.e., what limitations are placed on the interpretations that could be made)?

These experiments provide strong support for the hypothesis that biochemical and metabolic differences exist between metastasizing and nonmetastasizing cancer cells. These experiments proved highly productive because considerable care was exercised in making predictions and in forming additional hypotheses to identify where subsequent testing might be most likely to yield useful information. In the examples described here, the immune system and immunological reactions were examined in more detail since the body's immunological defense system seemed to provide a likely mechanism to prevent metastasis. While in a great many cases, experiments that seem to be the next logical step and potentially informative do not yield useful results, in this example the direction chosen for the research proved highly productive.

References

Feldman, M., and Eisenbach, L. November 1988. What makes a tumor cell metastatic? *Scientific American,* 60–85.

Fidler, I. J., and Kripke, M. L. August 26, 1977. Metastasis results from preexisting variant cells within a malignant tumor. *Science,* 893–95.

Case Study 8 METASTASIS OF CANCER CELLS

Name ____________________
Section ____________________
Date ____________________

I

Question 1—What possible alternative interpretation of the results from the original experiment does Result 1 eliminate?

Answer 1

II

Question 1—What if *no* difference in the H-2D and H-2K molecules was found in metastasizing and nonmetastasizing cells? Would this indicate that H-2D and H-2K are not involved in metastasis? Why or why not?

Answer 1

III

Question 1—Why was it important that the tumor cells *not* form tumors when injected into mice?

Answer 1

Question 2—Why couldn't this experiment have been performed by observing lymphocytes in the injected mice rather than the more complicated procedure of extracting lymphocytes and checking for interaction with cultured tumor cells?

Answer 2

IV

Question 1—Why was this experiment done? What additional information did it provide beyond Result 3?

Answer 1

Question 2—What is a primary deficiency of this experiment?

Answer 2

9 Case Study: The Golgi Apparatus (Difficulty Level 3)

The Golgi apparatus is an organelle that processes and packages proteins that will be used for specialized functions.

Experiments indicate that proteins synthesized at the endoplasmic reticulum enter the Golgi saccules at the forming face and exit at the maturing face (fig. 9.1). If proteins are modified as they pass through the Golgi, what would you expect to find in the Golgi that could bring about this modification?

Prediction 1—Enzymes that can modify the proteins should be found in the Golgi apparatus.

Result 1—When the Golgi apparatus is isolated from cells and purified, a wide variety of enzymes are found that would be able to modify the proteins. Some can add sugar molecules and others can add phosphate groups.

I

Why is finding an enzyme in association with a particular organelle a strong argument that a specific biochemical reaction occurs in the organelle?

If a specific enzyme is *not* found, does this mean that a certain reaction does not occur?

In another series of experiments, the Golgi saccules were separated from each other. This is accomplished by centrifuging the Golgi saccules in a sucrose density gradient in which cell fragments separate if these parts vary in density. Three groups (fractions) of the Golgi saccules were found. If each of these fractions contains saccules that carry out specific modifications, what would you expect to find?

Prediction 2—Specific enzymes will be found in the different fractions.

Result 2—Subsequent analysis showed that each of the 3 fractions contained specific enzymes.

II

In what ways could these data be misleading about Golgi structures?

Could the Golgi have more than 3 types of saccules? Fewer?

While these results suggest that specific reactions occur in different parts of the Golgi, they do not indicate exactly *where* the reactions are occurring. It is possible to determine where a reaction is located by doing this: (1) A purified Golgi enzyme is injected into a mouse so that the mouse's immune system manufactures antibodies to that enzyme. (2) A substance, which is visible under the electron microscope, is joined to the antibody molecules. (3) This modified antibody preparation, which will act as a stain, is then added to cells; the cells are sectioned and viewed under an electron microscope.

For each of the antibody preparations, would you expect to find only portions of the Golgi apparatus stained?

Prediction 3—The stain should be localized within specific saccules.

Result 3—Electron micrographs showed that each antibody preparation stained specific saccules. It became apparent that the Golgi apparatus has 3 compartments: an inner portion, a middle portion, and an outer portion.

III Summarize the data so far. Why do data show that "specific reactions occur in different parts of the Golgi?
What are the different parts?
Which of these hypotheses is best supported by the data already presented? Which, therefore, would you expect to be supported by further data?

You will recall that the Golgi apparatus has 2 faces, a forming face and a maturing face. It could be that proteins move through the Golgi in this manner: a saccule containing the protein gets pushed forward by new ones at the forming face. (Old saccules would constantly break down at the maturing face.) Or it could be that proteins move via the vesicles from one saccule to the next. This is called the "hopping" hypothesis because the vesicles would hop from the outer edge of one saccule to the next. (Notice that the figure shows vesicles at the edges of the saccules.) Which of these hypotheses is best supported by the data already presented? Which, therefore, would you expect to be supported by further data?

Figure 9.1 Golgi apparatus—it is hypothesized that material enters the apparatus at the forming face and exits in secretory vesicles. *(©Prof. Gordon Leedale/Biophoto Associates)*

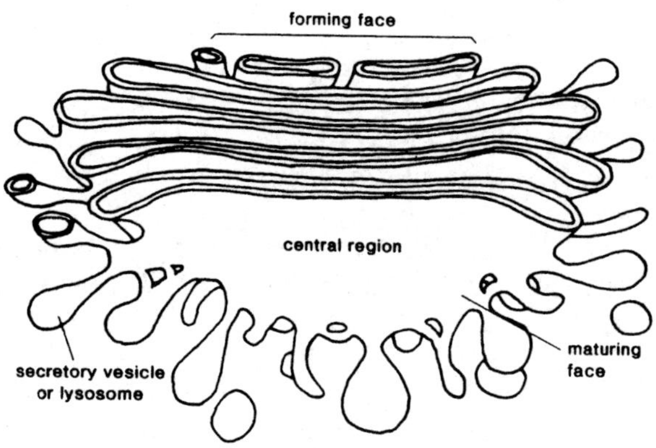

To test which of these 2 models accounts for the movement of substances through the Golgi, 2 cells were fused (i.e., plasma membranes were joined to produce one large cell containing contents of both original cells). One cell was a mutant cell and the other was normal. The mutant cell could produce protein G but could *not* add the sugar galactose to this protein. Prior to fusion, this cell was exposed to radioactive precursors so that all of the protein G in the Golgi was radioactively labeled. The normal cell could add galactose to protein G as it passes through the Golgi apparatus. The fused cell contained Golgi apparatus from both cells. Those from the mutant cell contained radioactive protein G (but could not add galactose) while Golgi from the normal cell could add galactose to the protein (but did not contain radioactive protein G). Knowing that vesicles would be able to move from one Golgi apparatus to the other, what would you expect to happen if the hopping model is correct?
Prediction 4—If hopping can occur (even between Golgi apparatuses), then radioactive protein G with galactose added should be found in the fused cell.
Result 4—Careful analysis indicated that radioactive protein G with galactose added was found. Therefore, the hopping model was supported.

IV In this test of the "hopping" hypothesis, why is it necessary to use cells that have a mutation preventing the addition of galactose?
What would happen if this mutation were not present?

These diverse experiments provide a more detailed description of the Golgi apparatus and indicate that there are several distinct biochemical compartments containing specific enzymes. Proteins appear to be transported from saccule to saccule via "hopping" of vesicles.

Reference

Rothman, J. E. September 1985. The compartmental organization of the Golgi apparatus. *Scientific American.*

Case Study 9 THE GOLGI APPARATUS

Name ____________________
Section ____________________
Date ____________________

I

Question 1—Why is finding an enzyme in association with a particular organelle a strong argument that a specific biochemical reaction occurs in the organelle?

Answer 1

Question 2—If a specific enzyme is *not* found, does this mean that a certain reaction does not occur?

Answer 2

II

Question 1—In what ways could these data be misleading about Golgi structures?

Answer 1

Question 2—Could the Golgi have more than 3 types of saccules? Fewer?

Answer 2

III

Question 1—Summarize the data so far. Why do data show that "specific reactions occur in different parts of the Golgi"? What are the different parts?

Answer 1

Question 2—Which of these hypotheses is best supported by the data already presented? Which, therefore, would you expect to be supported by further data?

Answer 2

IV

Question 1—In this test of the "hopping" hypothesis, why is it necessary to use cells that have a mutation preventing the addition of galactose? What would happen if this mutation were not present?

Answer 1

10 Case Study: Clonal Selection Theory (Difficulty Level 3)

When animals are exposed to foreign substances called antigens, they produce antibodies against them. Early investigators were aware of 2 hypotheses to explain the sudden appearance of antibodies in animals immunized to specific antigens (fig. 10.1):

Template hypothesis—Any antigen (and more than one type at a time) can enter any lymphocyte and once inside, an antigen acts as a *template,* causing a newly forming antibody to become specific. Then the antibody is secreted.

Clonal selection hypothesis—Each lymphocyte has one type of membrane-bounded antibody that is already specific. When an antigen combines with a membrane-bounded antibody, this lymphocyte is *selected* to give rise to a *clone* of plasma cells that secrete many specific antibodies.

Does each lymphocyte have the potential to produce more than one kind of antibody according to the template hypothesis? according to the clonal selection hypothesis? Is it possible for each lymphocyte to produce 2 or more kinds of antigens according to the template theory? according to the clonal selection hypothesis? Will there, with time, be many more cells to react to a specific antigen according to the template hypothesis? according to the clonal selection hypothesis?

Many experiments were done to test which of these 2 theories is supported by data. For example, in one experiment mice were immunized to 2 different types of antigens. The antigens were flagella proteins from 2 different motile bacteria. After a few days, the experimenter removed the lymph nodes from the mice, and one lymph node cell (lymph node cells are capable of producing antibodies) was placed per depression on a microscope slide. If a lymph node cell is producing antibody to a bacterial flagellum, movement of the bacterium ceases immediately. What do you predict would happen when both types of bacteria are added, one at a time to a depression containing a single lymph node cell according to the template hypothesis? according to the clonal selection hypothesis?

Prediction 1—If the template hypothesis is correct, both types of bacteria would be immobilized by each lymph node cell. If the clonal selection hypothesis is correct, only one type of bacterium would be immobilized by each lymph node cell.

Result 1—Only one type of bacterium was immobilized for each lymph node cell. Therefore, the clonal selection hypothesis was supported.

I If there had been more than one cell per depression what might have happened?
Would it have been possible to distinguish which hypothesis was supported?
Assuming that the clonal selection hypothesis is correct, what would have been the chance of an antibody-antigen reaction if lymph node cells had been taken from nonimmunized mice?

In another series of experiments, the spleen cells (where lymphocytes are also produced) of nonimmunized mice were exposed to a radioactively labeled antigen. The spleen cells were then washed free of any antigens that had not reacted to them. Which hypothesis—the template or clonal selection—would predict that very few spleen cells from nonimmunized animals would combine with the labeled antigen?

Prediction 2—Very few spleen cells from a nonimmunized animal would be found with bound antigen according to the clonal selection hypothesis.

Result 2—Subsequent measurement and analysis showed that, as predicted, only a few cells bound the antigen. These results are consistent with the clonal selection hypothesis.

Which hypothesis—the template or clonal selection—would predict that many cells will combine with the labeled antigen when spleen cells are taken from immunized mice?
Prediction 3—Many spleen cells from an immunized animal would bind to labeled antigen according to the clonal selection theory.
Result 3—Experiments showed that immunized animals have many spleen cells capable of binding with the labeled antigen.

Summarize the data from Results 1 through 3. How do the data fail to support the template hypothesis? On the other hand, how do the data support the clonal selection hypothesis?

According to the clonal selection hypothesis, a small number of antigen-binding cells, such as observed in Result 2, produce by division a larger number of antigen-binding cells, such as observed in Result 3.

II

Why was more antigen bound in the immunized mice? What results would you expect if the template hypothesis were supported?

Figure 10.1 The 2 hypotheses to explain the sudden appearance of antibodies in animals immunized to specific antigens.

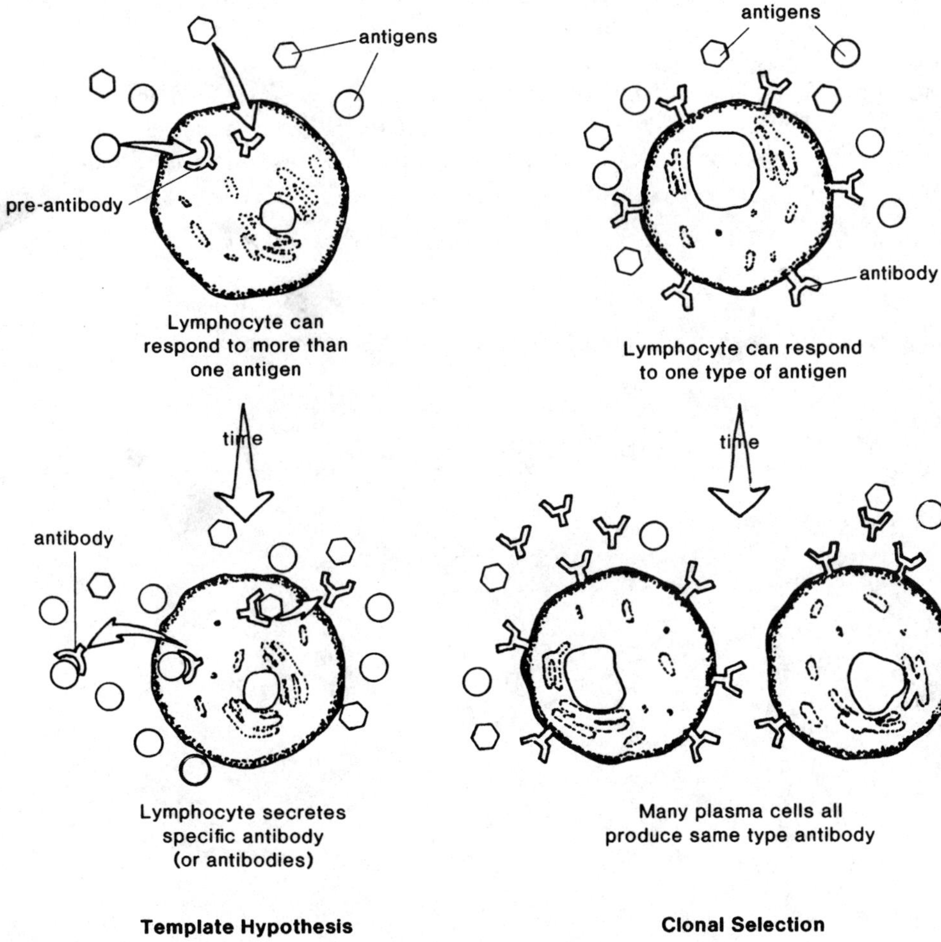

To observe this consequence, this experiment was performed: (1) Mice were injected with radioisotope-labeled antigen that destroyed any lymphocyte bound to the antigen. (2) All remaining lymphocytes were removed from these animals and (3) placed in mice whose immune systems had been destroyed by exposure to X rays. (4) These mice were injected with 2 types of antigens, a new one and the same one used previously, only this time neither antigen was labeled.

Predict which type antibodies the mice would produce if the clonal selection hypothesis is supported?

Prediction 4—If the clonal selection hypothesis is supported, the animals would produce antibody in response to the new antigen but not to the antigen used previously.

Result 4—As predicted, the mice could respond to the new antigen but not to the antigen that was used previously.

III Why were mice used for this experiment—why not just laboratory glassware?

Why do you need 2 groups of mice—one that can produce new lymphocytes and one that no longer can produce new lymphocytes? Why not just inject the antigen both times into the same animal?

Why do you need to use 2 different antigens? What was the role of the new antigen?

This series of investigations shows how a variety of experiments provide evidence to support one hypothesis over the other. Since experiments, such as those described here, gave increasing support to the clonal selection hypothesis of lymphocyte and antibody production, it is now called the clonal selection theory.

Reference

Gordon, L. A., and Nossal, G. August 1987. The clonal-selection theory. *Scientific American.*

Case Study 10 CLONAL SELECTION THEORY

Name ____________________
Section ____________________
Date ____________________

I

Question 1—If there had been more than one cell per depression what might have happened? Would it have been possible to distinguish which hypothesis was supported?

Answer 1

Question 2—Assuming that the clonal selection hypothesis is correct, what would have been the chance of an antibody-antigen reaction if lymph node cells had been taken from nonimmunized mice?

Answer 2

II

Question 1—Why was more antigen bound in the immunized mice?

Answer 1

Question 2—What results would you expect if the template hypothesis were supported?

Answer 2

III

Question 1—Why were mice used for this experiment—why not just laboratory glassware?

Answer 1

Question 2—Why do you need 2 groups of mice—one that can produce new lymphocytes and one that no longer can produce new lymphocytes? Why not just inject the antigen both times into the same animal?

Answer 2

Question 3—Why do you need to use 2 different antigens? What was the role of the new antigen?

Answer 3

PART II

Organism Structure and Function

11 Case Study: Gravitropism (Difficulty Level 1)

The ability of plant roots to orient to gravity has long fascinated casual observers as well as scientists. Roots are commonly observed to grow downward toward the earth. If a potted plant is placed on its side, the roots will bend and grow downward within a short time. This ability to detect gravity and orient to grow downward is termed *gravitropism* and is of obvious advantage to the plant. If roots are to grow into the soil and obtain the necessary water and nutrients, it is essential for the root to direct growth toward the earth.

How can roots detect the pull of gravity? How does the root manage to bend and grow in the appropriate direction? Before attempting to answer these questions, it will be necessary to examine the structure of the root.

Close observation of the root's cellular structure indicates several distinct regions. As shown in figure 11.1, the tip of the root is covered by a root cap. Behind the cap is the meristematic region, where cell division takes place and new cells are produced. In the zone of elongation, cells expand and elongate, accounting for the increase in the length of the root.

Investigations of gravitropism can be guided by the following hypothesis, which in turn leads to specific predictions about gravity detection and bending:

Hypothesis—Cells in a specific region of the root can detect gravity; if the root is not growing toward the earth, this detection will lead to a different growth rate between the top and the bottom of the root so the root will bend toward the earth.

To test the first part of the hypothesis—that cells in some part of the root can detect gravity—investigators removed different parts of the root and observed the effects on the gravitropic response. In one series of experiments, the root cap was removed and the plant placed in a horizontal position. What do you predict would be observed?

Prediction 1—If the root cap is the site where gravity is detected, the root will not bend downward when placed in the horizontal position.

Result 1—As shown in figure 11.2, roots that have had the root cap removed do not bend toward the earth but continue to grow straight. The untreated root in figure 11.2 shows the normal gravitropic bending response. Moreover, if the removed root cap is replaced on the tip of the root, the ability to exhibit the gravitropic response is regained. Thus, the root cap appears to be the part of the root that detects gravity. The identification of a specific site in the root where detection occurs supports the hypothesis that cells in a specific region of the root can detect gravity.

Question—Another proposed interpretation of Result 1 is that the damage produced when the root cap is removed abolishes the gravitropic response. This would mean that the root cap does *not* function to detect gravity. Is this a possible interpretation?

Answer—No. Replacement of the root cap restores the ability to respond to gravity. If the damage alone abolished the response, replacing the root cap would *not* have restored the response.

Question—What would be another way to test the possibility that damage alone abolishes the gravitropic response?

Answer—Intentionally damaging parts of the root by slashing it with a razor blade but without severing any part of the root and then observing the response. When this is done, the root still shows a normal gravitropic response, again demonstrating that damage alone does not affect the response.

If the root cap is removed from a root and the root cap from *another* root is placed on the cut stump, would you expect this root to show a gravitropic response?

What would happen if the root cap was removed and the root left in the vertical position?

More detailed observations of cells in the root cap have shown that many of the cells in the center part of the root cap contain relatively large organelles with a high starch concentration. These organelles, termed *amyloplasts,* are pulled to the bottom of the cell by gravity. When the root is reoriented, the amyloplasts can move in the cell. In other words, if the root is moved from the vertical to the horizontal position, the amyloplasts will move to the bottom side of the cell. It has been suggested that the amyloplasts are the cellular structures that detect gravity, but current research does not definitely indicate this is the case.

Figure 11.1 *a.* Structure of a typical root. *b.* Expanded view of central root cap cells showing amyloplasts.

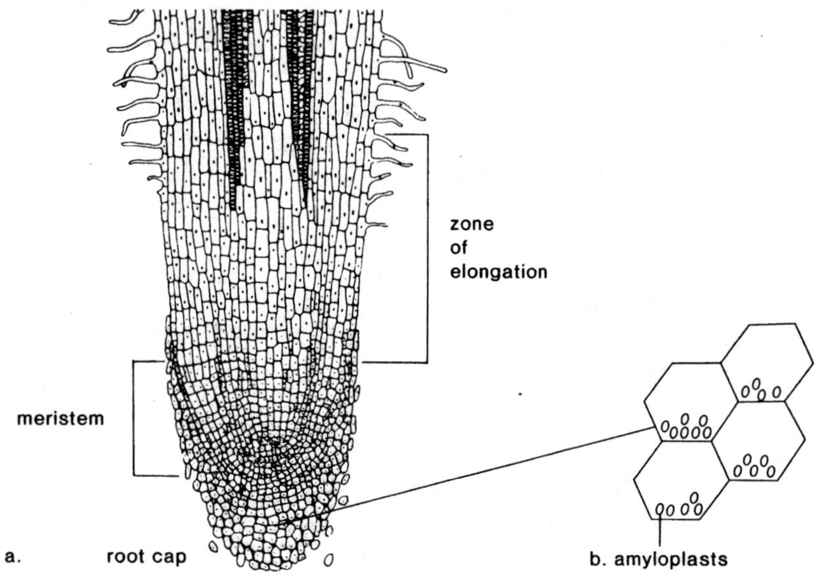

Figure 11.2 Root without root cap (*top*) continues to grow but does not bend. Intact root shows normal gravitropic response.

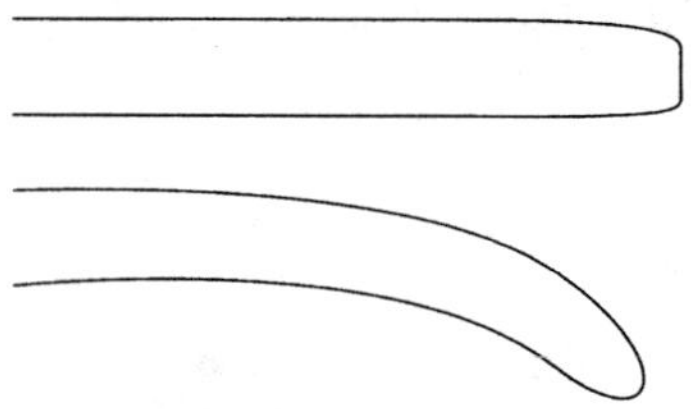

To test the second part of the hypothesis on the different growth rates between the top and bottom of the root, careful measurements were made of the growth rate on the 2 sides of the root.

A more accurate description might be the rate or extent of cell expansion, since the greatest increase in root length occurs because of cell expansion (fig. 11.3).

If a horizontal root is to bend downward, where would you expect the greatest growth or expansion to occur?

Prediction 2—If bending in the gravitropic response takes place because of different growth rates between the top and bottom of the root, the top of the root should grow faster than the bottom.

Result 2—Subsequent measurements indicated, as predicted, a faster growth rate on the top of the root as shown in figure 11.3. As expected, the maximum rate of growth (or expansion) occurred in the expansion zone about 3 mm from the tip of the root.

Question—If marks were placed close together on the surface of the root and the root allowed to grow for some period, where on the root would the marks be farthest apart?

Answer—Greatest separation should occur in the zone of elongation, where the cells are expanding. No increased separation should be observed on the root cap, where no expansion occurs.

II If the growth rates on the 2 sides of a *vertical* root are measured, what would be observed?
If cell elongation were inhibited, what would be observed in the gravitropic response?
How should a horizontal root be rotated to prevent the gravitropic response?

Figure 11.3 Gravitropic response of the root illustrating greater growth (cell expansion) on top of root leading to downward bending of the root.

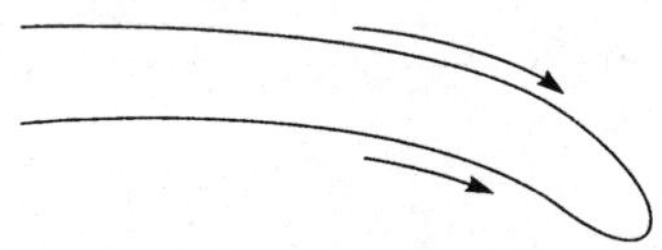

Another interesting observation was made while investigating the growth rates on the 2 sides of the root. During the bending response, the growth rate on *both* the top and the bottom sides of the root is *less* than the growth rate in the vertical root. In other words, the growth rate is decreased less on the top of the root than on the bottom. This observation suggests that the bending response is brought about by the release of some growth or cell expansion inhibitor, perhaps by the root cap when gravity is detected. To effect the bending response, the concentration of inhibitor would be greater on the bottom of the root than on the top.

Although plant hormones such as auxin or abscisic acid have been suggested as the possible growth inhibitor, clear evidence on the precise identity has not been obtained.

While the experiments described here provide strong support for the hypothesis, many questions remain. Does an inhibitor actually bring about the bending response? How could the shifting position of the amyloplasts release such an inhibitor? Continued investigations should reveal further evidence to help answer these questions.

Even a seemingly simple response, such as gravitropism, is comprised of many complex events. Investigation of these events requires a variety of experiments to examine the response in detail and to reveal the separate reactions that control and comprise the gravitropic response.

References

Evans, M. L., Moore, R., Hasenstein, K. H. December 1986. How roots respond to gravity. *Scientific American,* 112–19.

Moore, R., and Evans, M. L. April 1986. How roots perceive and respond to gravity. *American Journal of Botany,* 574–87.

Street, H. E., and Opik, H. 1976. *The Physiology of Flowering Plants,* 2d ed. London: William Clowes & Sons, Limited.

Case Study 11 GRAVITROPISM

Name ____________________
Section ____________________
Date ____________________

I

Question 1—If the root cap is removed from a root and the root cap from *another* root is placed on the cut stump, would you expect this root to show a gravitropic response?

Answer 1

Question 2—What would happen if the root cap was removed and the root left in the vertical position?

Answer 2

II

Question 1—If the growth rates on the 2 sides of a *vertical* root are measured, what would be observed?

Answer 1

Question 2—If cell elongation were inhibited, what would be observed in the gravitropic response?

Answer 2

Question 3—How should a horizontal root be rotated to prevent the gravitropic response?

Answer 3

Method 1

Method 2

Method 3

12 Case Study: Transport in Phloem (Difficulty Level 1)

The transport of organic nutrients in flowering plants takes place in the *phloem,* while water is moved in the xylem. Phloem cells, as shown in figure 12.1, are connected end-to-end to form tubular strands that conduct the organic material. Phloem tissue extends from the leaves through the stem to the roots and provides a continuous pathway for transport. Sugars produced in the leaves by photosynthesis can be transported downward to the roots, and other material can be moved in the upward direction as well.

Figure 12.1 Phloem tissue shown in the context of other plant cells (*a*) and as individual phloem cells (*b*).

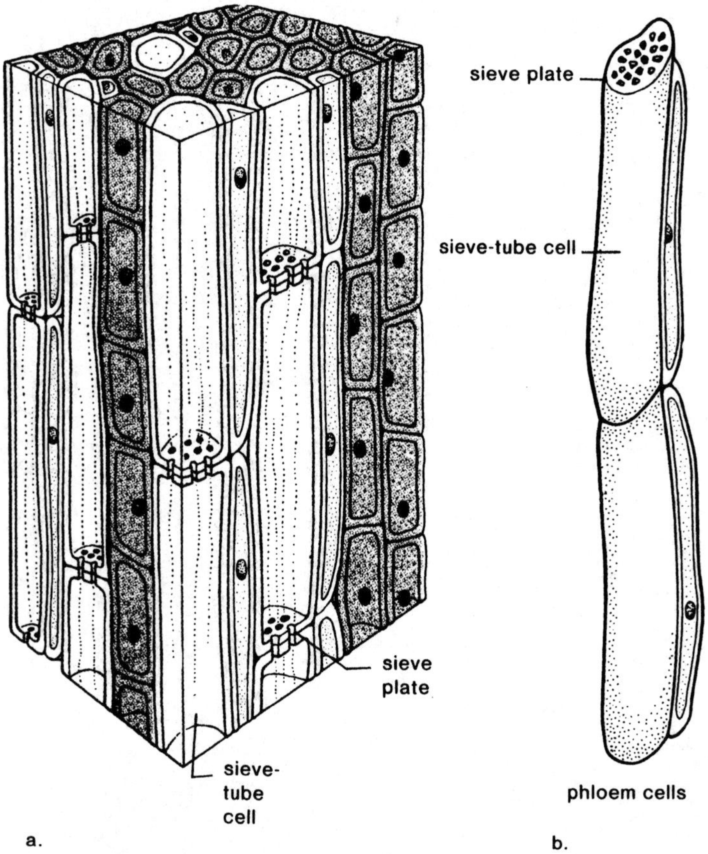

Plant physiologists have long wondered what causes transport in the phloem. Two major hypotheses have been proposed:

Hypothesis 1—Cytoplasmic streaming, the movement of cytoplasm in an orderly stream in one direction in the cells, can rapidly move material inside living cells. Organic material can be transported from one end of the phloem cell to the other by cytoplasmic streaming, cross the sieve plate, and continue to be transported in the same manner by the next phloem cell. Thus, rapid movement of large quantities of organic material through the phloem could occur.

Hypothesis 2—Sugar synthesized in the leaf by photosynthesis enters the phloem cells and increases the osmotic pressure. Water will then enter the leaf phloem cells in response to the increased osmotic pressure, increasing the water pressure, which forces the sugar and water through the phloem to the roots. This scheme, termed the *pressure-flow hypothesis,* is shown in figure 12.2. In this way, rapid bulk flow can occur in the phloem.

Figure 12.2 The pressure-flow hypothesis of phloem transport. *a.* A model system showing how increased sugar concentration at 1 causes water flow into the bulb and bulk flow through the system to 2. *b.* The proposed operation of pressure-flow in the phloem tissue.

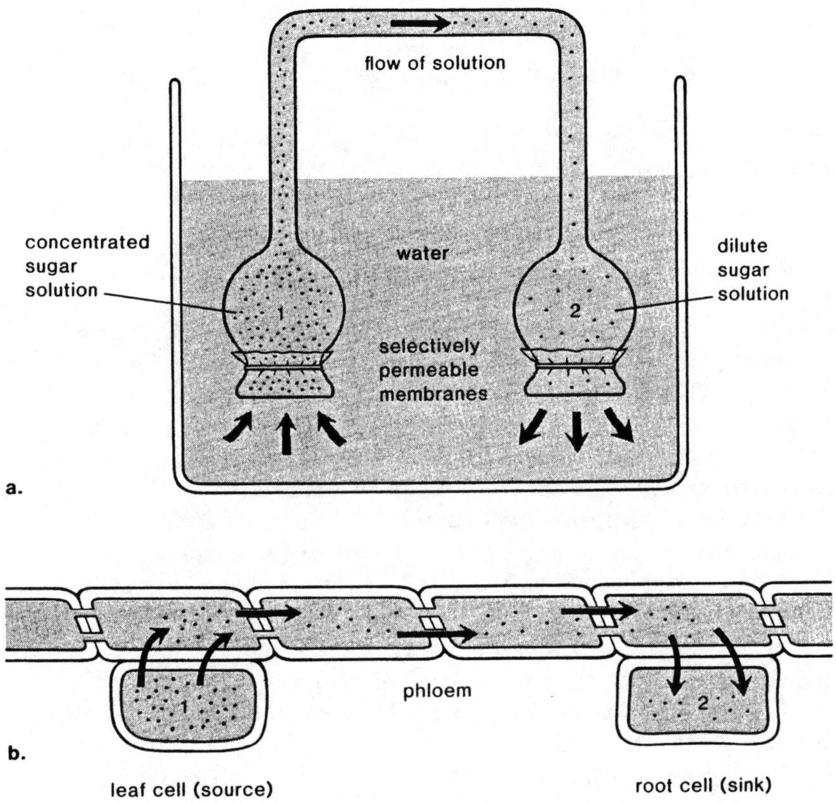

These 2 hypotheses represent 2 different yet reasonable explanations for phloem transport. But which of these is the more accurate description of what actually happens in the plant? To answer this question requires making predictions based on the hypotheses, performing experiments, and making observations to determine whether the predictions are consistent with the observations and results.

One major difference between the 2 hypotheses is that cytoplasmic streaming requires phloem cells to be alive for transport to take place. In the pressure-flow hypothesis, the phloem cells function more like "water pipes," serving simply as passive channels for water and sugar movement.

Is it necessary that phloem cells be alive when transport is taking place? Cytoplasmic streaming does stop when cells die, and several inhibitors slow down or stop streaming.

Prediction 1—If the cytoplasmic streaming hypothesis accounts for phloem transport, transport should stop when the phloem cells are killed.

Result 1—When a section of a plant stem is killed, transport through this section stops. This is in contrast to water transport that continues through the xylem. Xylem cells are dead at maturity yet continue to be passive tubes conducting water. This observation, therefore, is consistent with the cytoplasmic streaming hypothesis.
Question—Is there an alternative interpretation of Result 1 that is consistent with the pressure-flow hypothesis?
Answer—When the phloem cells are killed, perhaps the cytoplasm plugs the openings in the sieve plate thereby preventing flow through the sieve cells. In this way, flow predicted by the pressure-flow hypothesis would be stopped.

I

What effect would you predict on phloem transport if the rate of cytoplasmic streaming were slowed but not stopped?
Since an alternative interpretation of Result 1 can be made, how strongly does Result 1 *now* support the cytoplasmic streaming hypothesis?

Another question has been raised about the cytoplasmic streaming hypothesis. Is the rate of streaming fast enough to account for the rate of phloem transport. How fast can streaming transport material? Careful measurements made with a microscope while observing the streaming of cytoplasm indicate the streaming rate is approximately 0.04 cm per minute in a variety of cells. How does this compare with transport rates measured in phloem?
Prediction 2—If measurements of phloem transport are approximately equal or less than 0.04 cm per minute, these measurements would be consistent with both hypotheses. If the phloem transport rate is much higher, transport by cytoplasmic streaming would appear to be much less likely.
Result 2—The rate of movement in the phloem has been measured in a variety of ways. Common methods involve adding dyes or radioactive material to the phloem and observing how rapidly this material moves up the plant. One very interesting technique measured the rate of pumpkin growth and calculated the rate of sugar transport in the phloem necessary to maintain the observed rate of growth. All these methods have frequently indicated phloem transport rates of about 2 cm per minute with some as high as 8 cm per minute. These rates are far greater than could be explained by cytoplasmic streaming. Such results, therefore, suggest the pressure-flow hypothesis is the more likely explanation of the mechanism of phloem transport.
Question—How could the pressure-flow hypothesis explain the movement of material from the roots upward to the top of the plant as is sometimes observed?
Answer—If the sugar concentration entering the phloem at the roots is greater than the sugar entering at the leaves, more water will enter the root phloem, forcing movement toward the top of the plant.

II

What if measurements of the rate of phloem transport had been 0.5 cm per minute? This result would be consistent with which hypothesis?
The rate of phloem transport was measured in a variety of ways. How does this strengthen the validity of the result?
If the rate of photosynthesis was increased so more sugars were produced, how might this affect the rate of phloem transport as described by the pressure-flow hypothesis?

The observations and results on phloem transport provide stronger support for the pressure-flow hypothesis even though cytoplasmic streaming initially seemed to be a valid hypothesis.

This study is an interesting and informative one because it illustrates that 2 or more initial hypotheses can provide valid explanations, while subsequent experimentation and careful testing can narrow the explanation to one most likely hypothesis.

References

Street, H. E., and Opik, H. 1976. *The Physiology of Flowering Plants,* 2d ed. London: Edward Arnold Publishers Ltd.

Zimmerman, M. H. March 1963. How sap moves in trees. *Scientific American,* 132–42.

Case Study 12 TRANSPORT IN PHLOEM

Name ____________________
Section ____________________
Date ____________________

I

Question 1—What effect would you predict on phloem transport if the rate of cytoplasmic streaming were slowed but not stopped?

Answer 1

Question 2—Since an alternative interpretation of Result 1 can be made, how strongly does Result 1 *now* support the cytoplasmic streaming hypothesis?

Answer 2

II

Question 1—What if measurements of the rate of phloem transport had been 0.5 cm per minute? This observation would be consistent with which hypothesis?

Answer 1

Question 2—The rate of phloem transport was measured in a variety of ways. How does this strengthen the validity of the result?

Answer 2

Question 3—If the rate of photosynthesis was increased so more sugars were produced, how might this affect the rate of phloem transport as described by the pressure-flow hypothesis?

Answer 3

13 Case Study: Photosynthesis and Oxygen Production (Difficulty Level 1)

The investigation of photosynthesis using aquatic plants is one of the most common experiments performed in introductory biology laboratories. In these experiments, *Elodea* is the plant most often used and the rate of photosynthesis is estimated by the amount of oxygen produced. Since *Elodea* is an aquatic plant, oxygen can be measured as the number of bubbles liberated into the water by the plant or by collecting and measuring the total volume of gas produced.

Using this experimental arrangement, the effects of several variables on photosynthesis can be investigated. The overall summary equation of photosynthesis,

Equation 1

$$6CO_2 + 6H_2O \xrightarrow{\text{light energy}} 6O_2 + C_6H_{12}O_2 \text{ (sugar)}$$

predicts that as the plant is exposed to more light and to increasing concentrations of CO_2, more O_2 and sugar are produced. The effects of several other variables such as temperature, color of light, and pH (acidity) of the water bathing the plant have also been investigated. Especially interesting results have been obtained from experiments on pH effects, as shown in figure 13.1, where a much greater volume of gas is produced at pH 5 than at pH 8. Why should this effect of pH be observed?

Figure 13.1 Gas production in *Elodea* at pH 5 and 8.2.

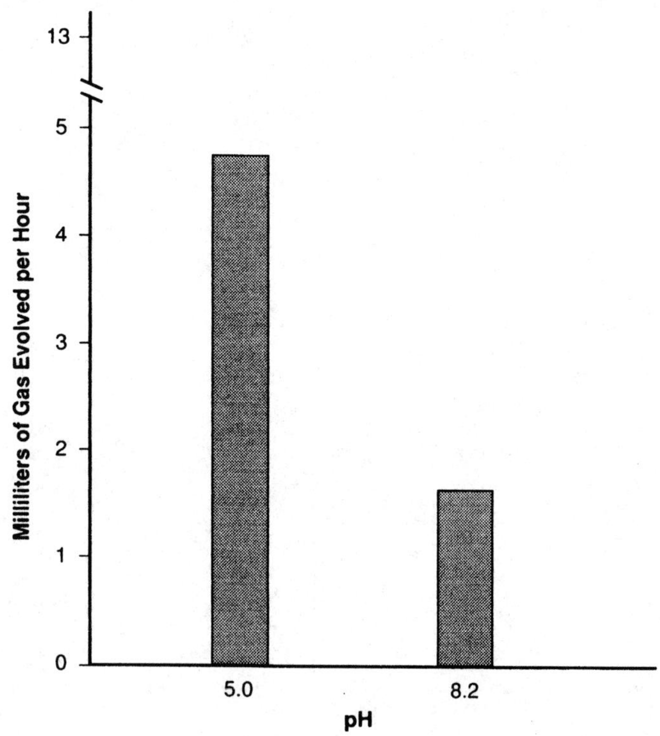

Several hypotheses could be proposed, but perhaps the most logical is the following:

Hypothesis—At acid pH levels (i.e., pH 5), CO_2 is at a greater concentration in the solution. This increased CO_2 results in a greater rate of photosynthesis as indicated by equation 1.

This is a logical hypothesis because pH is known to affect the amount of CO_2 in the solution. Would you predict that low pH would increase or decrease CO_2 in the solution?

Prediction 1—Acid pH levels should *increase* the amount of CO_2 in the solution. The increased CO_2 would account for the greater rate of photosynthesis at pH 5.

Result 1—As shown in table 13.1, the amount of CO_2 in the solution increases as the solution becomes more acidic. This result clearly supports the hypothesis.

Table 13.1

pH	**Free CO_2 in Solution** (grams CO_2/liter)
7	0.3
6	0.9
5	1.5

Question—What is another hypothesis that would explain the effects of pH?

Answer—Possibly the low pH increases the activity of enzymes involved in photosynthetic reactions so that the reactions proceed at a faster rate, increasing the rate of photosynthesis. This hypothesis is, however, much more difficult to test; it would require isolating the enzymes and investigating the effect of pH on reaction rate. Also, Result 1 provides strong apparent support for the original hypothesis.

What experimental conditions of pH and light should result in the greatest rate of photosynthesis?

Is it possible that pH affects CO_2 concentration *and* has some other direct effect on the plant?

In another more detailed experiment, the amount of oxygen produced by *Elodea* was measured with the surrounding water pH at *several* different values between pH 5 and pH 8.2 under both light and dark conditions. The pH of the water bathing the *Elodea* was adjusted to the desired pH by adding sulfuric acid. The total amount of gas produced by the plant was collected and reported as milliliters of gas per hour. At each pH, measurements were made in constant light and in the dark. Considering Result 1, what results would you predict?

Prediction 2—Since more CO_2 is in the solution at acid pH values, more oxygen should be produced at pH 5 than at any of the more basic pH values. No oxygen should be produced in the dark at any of the pH values.

Result 2—As shown in figure 13.2, more gas was produced at pH 5 than at more basic pH values. There was an entirely unexpected result, however; a considerable amount of gas was produced in the dark! Since no photosynthesis (and no oxygen production) can occur in the dark, how is this gas produced?

Figure 13.2 Gas production in *Elodea* at several pH values in both light and dark conditions. The pH values indicate the pH of the water surrounding the *Elodea* plant.

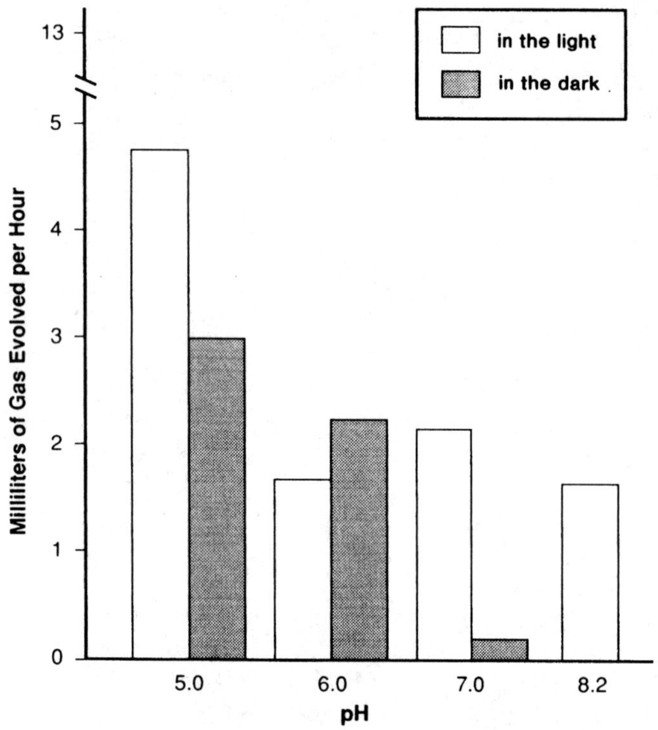

An obvious question to ask at this point is: What *is* the gas being produced? Is the gas produced in the dark actually oxygen or some other substance? These questions were investigated by analyzing the composition of the gas, and the results are shown in table 13.2.

Table 13.2 Gas Composition

	Treatment	% O_2	%CO_2	%N_2
Light	pH 8.2	47.3	1.4	36.3
	pH 6.0	21.8	40.9	18.5
Dark	pH 8.2	—	—	—
	pH 6.0	16.3	40.8	29.2

One immediate observation is that a large amount of CO_2 is produced in the light *and* dark at pH 6. At pH 8, however, very little CO_2 is produced. Instead, the largest component of the gas is O_2. Apparently some reaction is taking place at low (i.e., acid) pH values that produces CO_2, but this reaction does not occur at higher (i.e., basic) pH values. Therefore, experiments to investigate photosynthesis in *Elodea* by measuring oxygen production should be conducted at pH values near 8.

Question—If Result 1 clearly supported the hypothesis, why was a second experiment performed? Didn't the first experiment prove the hypothesis?

Answer—First, a hypothesis is never "proven" for sure. It is only supported by results that agree with predictions drawn from the hypothesis. Moreover, the greater the number of *different* predictions and results found that support the hypothesis, the greater our confidence that the hypothesis is "true" (but one can never be absolutely sure). Thus, the second experiment represented an attempt to increase confidence in the hypothesis. As it turned out, Result 2 revealed an unexpected effect that allowed a more complete understanding of *Elodea* physiology.

Question—What could account for the CO_2 production at low pH values?

Answer—One possibility is that a process or enzyme that produces CO_2 is much more active at low pH values than at high pH values. One such enzyme is carbonic anhydrase, an enzyme that catalyzes the conversion of carbonic acid (H_2CO_3) to CO_2 and water. Apparently, this enzyme is more active at low pH values.

II

What was the control in the experiment that yielded the results in figure 13.2?
Why was the control especially informative in this experiment?
Why is the gas production so high at pH 5 in the light in figures 13.1 and 13.2?
How would the results in figure 13.2 be interpreted if no measurements had been made in the dark?

These investigations illustrate 2 very important points:

1. It is important to test a hypothesis with several predictions and experiments. As shown in Result 1, a single prediction and test can support the hypothesis, but further tests can reveal that support is not as strong as initially expected.
2. Experimental results can show effects that are unexpected and surprising, and these results can in turn lead to new and more informative investigations.

While these results *do* provide support for the hypothesis, other effects are revealed that indicate that the hypothesis cannot completely account for the observations. Additional experiments are necessary to account for these unexpected results. Is the increased CO_2 production due to the action of carbonic anhydrase? Can the effects of carbonic anhydrase be separated from the photosynthetic response? The listed reference describes several possible investigations that address these questions.

Reference

Kosinski, R. J., and Stroup, D. J. May 1987. How-to-do-it, a demonstration for your lab: *In vivo* carbonic anhydrase activity in *Elodea. The American Biology Teacher,* 293–97.

Case Study 13 PHOTOSYNTHESIS AND OXYGEN PRODUCTION

Name ____________________

Section ____________________

Date ____________________

I

Question 1—What experimental conditions of pH and light should result in the greatest rate of photosynthesis?

Answer 1

Question 2—Is it possible that pH affects CO_2 concentration *and* has some other direct effect on the plant?

Answer 2

II

Question 1—What was the control in the experiment that yielded the results in figure 13.2?

Answer 1

Question 2—Why was the control especially informative in this experiment?

Answer 2

Question 3—Why is the gas production so high at pH 5 in the light in figures 13.1 and 13.2?

Answer 3

Question 4—How would the results in figure 13.2 be interpreted if no measurements had been made in the dark?

Answer 4

14 Case Study: Developmental Polymorphism (Difficulty Level 1)

Some organisms can develop into very dissimilar forms because of differences in environmental conditions. Such varied forms are termed *developmental polymorphisms.* One of the most impressive examples is found in the caterpillars of a moth (*Nemoria arizonaria*) found in the southwestern United States. The moth lays eggs only on trees of the Arizona oak. The caterpillars eat only the leaves and flowers of the oak tree. Although all young caterpillars appear identical must after they hatch, those hatching in the early spring grow to resemble closely the flowers of the tree (catkins), while those hatching later in the summer resemble the twigs of the tree. These different forms are shown in figure 14.1. One advantage of developing into these forms is immediately obvious when examining the figures. The camouflage provided by the forms should allow the caterpillars to avoid predators more effectively.

What causes the caterpillars to develop into these different forms? If some environmental factor is responsible (as would be expected with a developmental polymorphism), what is the factor? Since the different forms are observed in the early spring (when only the catkin-resembling form develops) and the summer (when only the twig-resembling form develops), it is logical to propose that any controlling factor would be an environmental variable that is itself different between early spring and summer. This proposal can be stated as follows:

Hypothesis—An environmental variable that is different between early spring and summer causes the caterpillar to develop into different forms.

What environmental variables would be most likely to control this developmental process?

Two such variables can be suggested. First, a variation in temperature is the most obvious. Second, it has been observed that early spring caterpillars feed almost exclusively on the oak flowers (shown in fig. 14.1), while in summer when flowers are no longer present, caterpillars feed solely on leaves.

How could the effect of these variables of temperature and food on the caterpillar's body be tested? In designing such an experiment, it is important to test all possible combinations of food and temperature. In this experiment, the following combinations are possible: early spring temperature and catkins, early spring temperature and leaves, summer temperature and catkins, summer temperature and leaves.

Another way of presenting these combinations is as follows:

Temperature	Food	
	Catkins	Leaves
Early spring		
Summer		

All combinations should be tested to determine whether both factors or only a single factor have an effect.

Eggs from a single female moth were divided into 4 groups. Young caterpillars were raised at temperatures characteristic of early spring (15°C) or summer (25°C) and were fed either catkins or leaves.

Figure 14.1 Different developmental forms of the caterpillar *Nemoria arizonaria. a.* Early spring form on a catkin. The caterpillar closely resembles the catkins. *b.* The twig-resembling form of the caterpillar appearing later in the summer. *(©Dr. Erick Greene)*

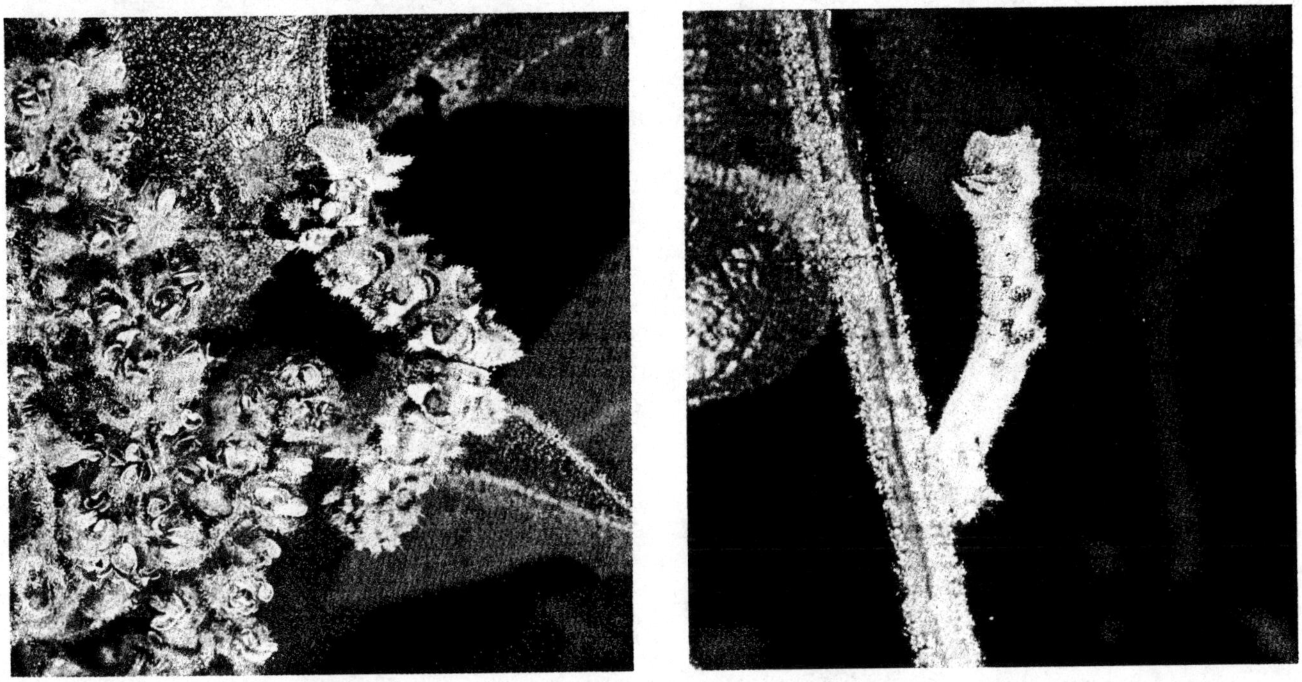

a. b.

Prediction 1—If food only determines caterpillar form, then temperature will have no effect. But if only temperature determines form, then food will have no effect.
Result 1—The results obtained in this experiment are shown in table 14.1.

Table 14.1

Temperature	**Food**	
	Catkins	Leaves
Early spring (15°C)	all catkin forms	all twig forms
Summer (25°C)	all catkin forms	all twig forms

The entries indicate the form of caterpillar development after each treatment.

Temperatures had *no* effect on the caterpillars. In other words, all caterpillars fed catkins developed into catkin-resembling forms, and all those fed leaves developed into twig-resembling forms.
Question—Why were eggs from only a single female moth used?
Answer—Using eggs from a single parent eliminates any variability due to genetic differences in separate individuals. If eggs from more than one female were used, then any observed differences in caterpillar form *might* have been due to genetic differences between the females and not due to environmental variables.

Question—If *both* food and temperature affected caterpillar form, what results would have been observed?
Answer—You would expect to find variability in form with both temperature and food. For example, some of the caterpillars fed catkins and raised at 25°C would have developed into twig forms, while all those fed catkins and raised at 15°C would have developed into catkin forms.

I Based on the results of this experiment, what would you expect to observe with caterpillars raised at 20°C and fed only catkins?
Why would feeding all groups of caterpillars both catkins *and* leaves be a poor experiment?

In a second experiment, the effects of the food consumed by caterpillars were examined in more detail. Catkins are known to be high in protein and low in tannin (an organic compound that is poisonous to many animals). Leaves, on the other hand, are relatively low in protein but high in tannin. Do proteins and/or tannins affect caterpillar form?

This question was examined by feeding separate groups of young caterpillars the following foods:

Diet	Food
1.	Catkins only
2.	Catkins + leaves
3.	Catkins + added tannins
4.	Leaves only

You would expect caterpillars fed catkins only to develop into catkin-resembling forms and those fed leaves only to develop into twig-resembling forms.

What would you expect to observe with caterpillars fed the other diets?
Prediction 2—If tannins cause caterpillars to develop into twig forms, then diets 2 and 3 should cause development into twig forms.
Result 2—The results of this experiment are shown in table 14.2

Table 14.2 The Effect of Diet on Developmental Form

Diet	Food	% Twig Forms
1.	Catkins only	6%
2.	Catkins + leaves	88%
3.	Catkins + added tannins	94%
4.	Leaves only	94%

These results indicate that tannins are the substances that cause development of the twig forms. Apparently, it is the lack of significant amounts of tannin in catkins that leads to the development of catkin-resembling forms.
Question—If proteins *caused* development into catkin forms, what would you expect to observe with caterpillars fed diets 2 and 3?
Answer—These caterpillars would then be expected to develop into catkin-resembling forms since relatively high protein would be provided by the catkins.

II If young caterpillars were fed leaves + tannins, what form would develop?
If young caterpillars were fed catkins + leaves and temperature was varied, what would you expect to observe?

These experiments show how testing the effects of different variables (i.e., temperature and diet) can indicate what is controlling development. An extension of the first experiment provided additional information on the component in the food source that exerted the greatest effect.

Further questions could be asked and investigated. For example: Do moths developing from catkin-resembling caterpillars have a survival advantage? Do tannins affect the caterpillars in ways other than form? If so, how?

Thus, from an initial observation on diverse caterpillar form, a variety of questions and experiments can be pursued.

Reference

Greene, E. February 3, 1989. A diet-induced developmental polymorphism in a caterpillar. *Science,* 643–46.

Case Study 14 DEVELOPMENTAL POLYMORPHISM

Name ____________________
Section ____________________
Date ____________________

I

Question 1—Based on the results of this experiment, what would you expect to observe with caterpillars raised at 20°C and fed only catkins?

Answer 1

Question 2—Why would feeding all groups of caterpillars both catkins *and* leaves be a poor experiment?

Answer 2

II

Question 1—If young caterpillars were fed leaves + tannins, what form would develop?

Answer 1

Question 2—If young caterpillars were fed catkins + leaves and temperature was varied, what would you expect to observe?

Answer 2

15 Case Study: Chemical Defenses of Plants (Difficulty Level 2)

Plants appear to be at a considerable disadvantage in their struggle for survival against herbivores. Herbivores can actively search out and consume plants, but the plants can neither evade the herbivores nor actively fight against such attacks. Are plants completely defenseless? What mechanisms could be used to provide an effective defense?

One clue to a potential defense is the impressive variety of chemicals that plants can synthesize. Chemicals have the potential to repel herbivores by functioning as poisons, as repellants, and as chemicals capable of disrupting the normal growth cycle of an animal. An obvious hypothesis is as follows:

Hypothesis—Chemicals produced by plants can provide an effective defense against predators.

One experiment to test this hypothesis was performed with caterpillars of the black swallowtail butterfly. These caterpillars feed voraciously on plants of the *Umbelliferae,* which includes plants such as celery and parsley. On the other hand, the *Cruciferae,* which includes the mustards, are completely avoided by the caterpillars. It is known that the *Cruciferae* contain a toxic compound termed *sinigrin.*

Is sinigrin harmful to the black swallowtail caterpillars? To investigate this question, celery leaves were treated so that sinigrin was absorbed, and the caterpillars fed on the treated leaves.

What do you predict would be observed with these caterpillars?

Prediction 1—If sinigrin is poisonous to black swallowtail caterpillars, they should not survive after eating leaves containing this compound.

Result 1—The celery leaves that contained as much sinigrin as is normally found in mustard leaves were lethal to the caterpillars. This experiment indicates that a chemical produced by a plant can serve as a defense against a herbivore.

Question—Sinigrin could have been added to an artificial diet and fed to the caterpillars. Would this have been a better experiment than feeding the caterpillars celery leaves that contained sinigrin?

Answer—No. Since celery leaves are the usual diet of the caterpillars, the added sinigrin is the only variable in this treatment. By using the celery leaves, this experiment shows that the normally acceptable host plant can be rendered harmful when a chemical is added. If some other artificial diet were used, the caterpillars would be subjected to other variables besides sinigrin.

I What would be observed if caterpillars were fed an artificial diet that included sinigrin?

Poisons are only one class of chemical compounds that could provide a defense for plants. The growth of insects, for example, is regulated by growth hormones as development proceeds through the stages of larvae, pupae, and adult. Are such hormones or hormonelike substances found in plants? Is there any evidence they can provide a defense mechanism?

Prediction 2—If growth hormone chemicals are found in plants, these plants should be protected from predation by insects.

Result 2—Several hormonelike substances have been found in plants that escape insect predation. One of the most striking examples is the bugleweed that grows in eastern Africa. It has been observed that in areas where locust swarms consumed almost all the vegetation, bugleweed was *not* attacked by the insects. When extracts of bugleweed were fed to insects, the larvae were prevented from shedding their exoskeleton during growth and could not survive. More detailed chemical analysis of the extract revealed several substances very similar to insect growth hormones, which interfere with the normal metamorphosis from larvae to pupae. Thus, this plant appears to achieve a considerable measure of protection due to the presence of the hormonelike chemical.

II Why was an extract fed to the insects that were tested? Why not just feed them bugleweed leaves and observe the insects' responses?

If you observed one species of plant that was not being preyed upon by insects, while other different, nearby species *were* being consumed, why could you *not* definitely conclude the first species had a defense against insects?

If plants can evolve chemical defense mechanisms against insect herbivores, perhaps insects can evolve countermeasures against the plants' defenses. This possibility can be stated as the hypothesis that some insects consume plants that have chemical defenses without being harmed. Observations that contributed to testing this hypothesis have revealed unexpected and interesting relationships between plants, some species of insects, and predators of the insects. For example, if blue jays consume monarch butterflies that have fed on milkweed plants, the birds become violently ill. The particular species of milkweed on which the butterfly feeds is known to contain high concentrations of cardenolides. Cardenolides are toxic chemicals that have a harmful effect on normal heart function in animals.

What appears to be happening in this interaction between the milkweed, monarch butterfly, and blue jay?

Prediction 3—The butterfly is able to store the cardenolides from the milkweed. When the blue jay consumes the butterfly, these chemicals make the blue jay ill.

Result 3—When the chemical components of the monarch butterfly's body are analyzed, cardenolides are found to be present in amounts sufficient to kill a small mammal. This result clearly provides strong support for the hypothesis. It is no surprise that the blue jay becomes violently ill. A large number of other insect-eating bird species avoid eating monarch butterflies. It is observed also that butterflies raised on plants other than milkweed have no harmful effects on the blue jays when consumed. Thus, the monarch butterfly has apparently used the chemical defenses of a plant to its own advantage.

III Why is the observation that "butterflies raised on plants other than milkweed have no harmful effects on the blue jays when consumed" an important one?

What might be concluded if this observation were not made?

Why don't the cardenolides harm the butterfly? (Propose at least one hypothesis that could be tested.)

The evolution of chemical defense mechanisms in plants and the counterresponse of the evolution by animals of ways to overcome or utilize these defensive chemicals may be examples of co-evolution. (See Case Study 21, "Co-Evolution.") Plants and animals can be in an "evolutionary arms race"—each constantly being acted on by natural selection to overcome the defenses of the other.

Reference

Rosenthal, G. A. January 1986. The chemical defenses of higher plants. *Scientific American,* 94–99.

Case Study 15 CHEMICAL DEFENSES OF PLANTS

Name ____________________
Section ____________________
Date ____________________

I

Question 1—What would be observed if caterpillars were fed an artificial diet that included sinigrin?

Answer 1

II

Question 1—Why was an extract fed to the insects that were tested? Why not just feed them bugleweed leaves and observe the insects' responses?

Answer 1

Question 2—If you observed one species of plant that was not being preyed upon by insects, while other different, nearby species *were* being consumed, why could you *not* definitely conclude the first species had a defense against insects?

Answer 2

III

Question 1—Why is the observation that "butterflies raised on plants other than milkweed have no harmful effects on the blue jays when consumed" an important one?

Answer 1

Question 2—What might be concluded if this observation were not made?

Answer 2

Question 3—Why don't the cardenolides harm the butterfly? (Propose at least one hypothesis that could be tested.)

Answer 3
Hypothesis:

Hypothesis:

Hypothesis:

16 Case Study: Water Transport in Xylem (Difficulty Level 2)

How does water reach the top of tall plants, such as trees? The cohesion-tension model of xylem transport has been hypothesized as follows:

Hypothesis—The cohesion-tension model of xylem transport is a mechanism to explain transport water to the top of the tallest tree.

This hypothesis proposes that the evaporation of water at the leaves pulls water through the xylem to the top of the plant. Enormous force must be exerted on the column of water inside the plant if the water is being "pulled" through the xylem.

Is the column of water strong enough to withstand these forces? Or, will the column be pulled apart at lower forces?

Prediction 1—The force necessary to "pull apart" a column of water will be greater than the force necessary to pull a column of water to the top of a tall tree.

Result 1—By spinning a glass capillary tube in a centrifuge, very large forces of tension could be exerted on the column of water in the capillary tube. Measurements of these forces indicate that water columns are not pulled apart until subjected to forces *10 times* those necessary to pull water to the top of a tall tree.

Question—Does this result *prove* that cohesion-tension accounts for water movement through the xylem?

Answer—No, but the results do indicate that water columns could withstand the forces necessary to transport water and are, thus, consistent with the hypothesis. If it had been found that the water column was pulled apart at much lower forces, then the hypothesis would have a serious weakness.

A stronger test of the hypothesis requires a direct measure of the tension, or "pull," on water columns in the xylem of plants. These measurements could be compared to estimates of the pull necessary to move water.

Investigators searching for a technique to measure the pull on xylem water developed a pressure bomb apparatus (fig. 16.1). A cut stem from a tree is placed in the pressure tank and sealed so that the cut end extends outside the tank. Gas is injected into the tank at increasing pressure until water is forced out of the cut end of the stem. (When the stem is first cut, water in the xylem at the cut appears to be pulled farther up into the stem.) The pressure necessary to force water from the stem balances the pull on water in the xylem and is a measure of the tension in the xylem water.

Figure 16.1 Pressure-bomb apparatus.

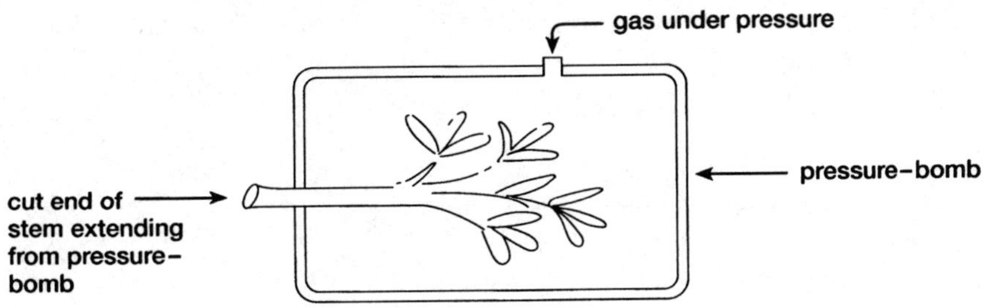

The pull necessary to move water to the top of a 100 m redwood is estimated to be about -20 atmospheres (-1 atmosphere would lift water about 10 m).

What would you predict about the pull measured with the pressure-bomb apparatus?

Prediction 2—A pull will be measured sufficient to move water to the top of a tree if cohesion-tension is moving water in the xylem.

Result 2—Forces in excess of -20 atmospheres have been measured in stems from the top of trees less than 100 m tall. Since this force is clearly sufficient to move water to the treetop, these measurements support the cohesion-tension hypothesis.

I What would be observed if the pressure in the bomb were increased well beyond the point that balances the pull on xylem water?

What would happen if evaporation of water from the leaves was decreased?

In additional measurements, stems from different heights on the same tree were used for measurements in the pressure-bomb apparatus.

What results would be predicted in these measurements?

Prediction 3—Forces in stems higher on the tree should be higher than forces measured from stems lower on the tree.

Result 3—Table 16.1 presents data obtained from one tree.

Table 16.1

Height on Tree	Measured Force
82 m	-17 atmospheres
57 m	-14 atmospheres
15 m	- 9 atmospheres

These data are consistent with the prediction and in every case are sufficient to pull water to the indicated height.

II Salt in soil water can exert a counterforce, tending to move water from a plant's roots to the soil if the plant is growing in salty water.

What would you expect to find if pressure-bomb measurements were made on such plants?

Reference

Scholander, P. F., et al. Sap pressure in vascular plants. *Science* 148: 339–46.

Case Study 16 WATER TRANSPORT IN XYLEM

Name ______________________
Section ______________________
Date ______________________

I

Question 1—What would be observed if the pressure in the bomb were increased well beyond the point that balances the pull on xylem water?

Answer 1

Question 2—What would happen if evaporation of water from the leaves was decreased?

Answer 2

II

Question 1—Salt in soil water can exert a counterforce, tending to move water from a plant's roots to the soil if the plant is growing in salty water. What would you expect to find if pressure-bomb measurements were made on such plants?

Answer 1

17 Case Study: The Causes of Cancer (Difficulty Level 2)

Almost 20% of the deaths in the United States are due to cancer. People tend to view this disease with more concern and fear than most other diseases because the illness can be painful and prolonged, as well as difficult and expensive to treat.

In spite of this concern, investigating the causes of cancer has been a difficult endeavor. It does not appear to be a simple infection by a disease-causing organism, nor does it have easily understood physiological causes. It is known that cancers are initiated by the uncontrolled division of groups of cells. These may arise anywhere in the body, and in many cases cancerous cells can spread throughout the body, giving rise to many secondary growths.

What causes normal cells to begin uncontrolled division and become cancerous? Two potential causes have been suggested by investigators:

1. Exposure to some environmental factors may cause cells to become cancerous.
2. The tendency to develop cancer may have genetic causes and could be inherited as are certain genetic diseases.

Most recently, environmental factors such as smoking and diet have attracted attention as possible cancer-causing agents and lead to the following hypothesis:

Hypothesis—Certain environmental factors cause cells to become cancerous.

When investigating this hypothesis, the possibility should not be neglected that genetics may be involved to some extent. Interesting studies have been made with identical twins to investigate if these individuals (who are genetically identical) would tend to show the same frequency and kinds of cancer.

What would you predict?

Prediction 1—If environmental factors cause cancer and genetic factors do not, comparisons of cancer rates in genetically identical individuals (identical twins) would show no similarities.

Result 1—Extensive comparisons of kinds and frequencies of cancer in nonidentical and identical twins have indicated no similarities. In other words, comparisons between pairs of nonidentical twins showed just as many differences in kinds and frequencies of cancer as comparisons between identical twins.

Question—Why were identical twins compared to nonidentical twins? Why not compare pairs of identical twins to pairs of completely unrelated individuals?

Answer—The investigation was to examine the effects of genetics on cancer with as little effect by other variables as possible. Comparisons between identical twins (no genetic differences) to those between nonidentical twins (some genetic differences) allows comparison between groups with relatively well-known genetic differences. Moreover, twins are raised in very similar environmental conditions, at least in their younger years. Unrelated individuals, however, would likely be raised in different environments, with significant variations in environmental factors that could account for differences in cancer rates.

If genetic factors show no significant relation to cancer, what about environmental factors? One of the most detailed studies is on the relation between smoking and lung cancer.

If an environmental factor (cigarette smoke) causes cancer, what would you predict?

Prediction 2—The greater the exposure to tobacco smoke, the greater the incidence of lung cancer.

Result 2—Figure 17.1 shows that as cigarette consumption increases, so does the incidence of cancer. Similar trends are observed for men and women.

Figure 17.1 Smoking and incidence of lung cancer.

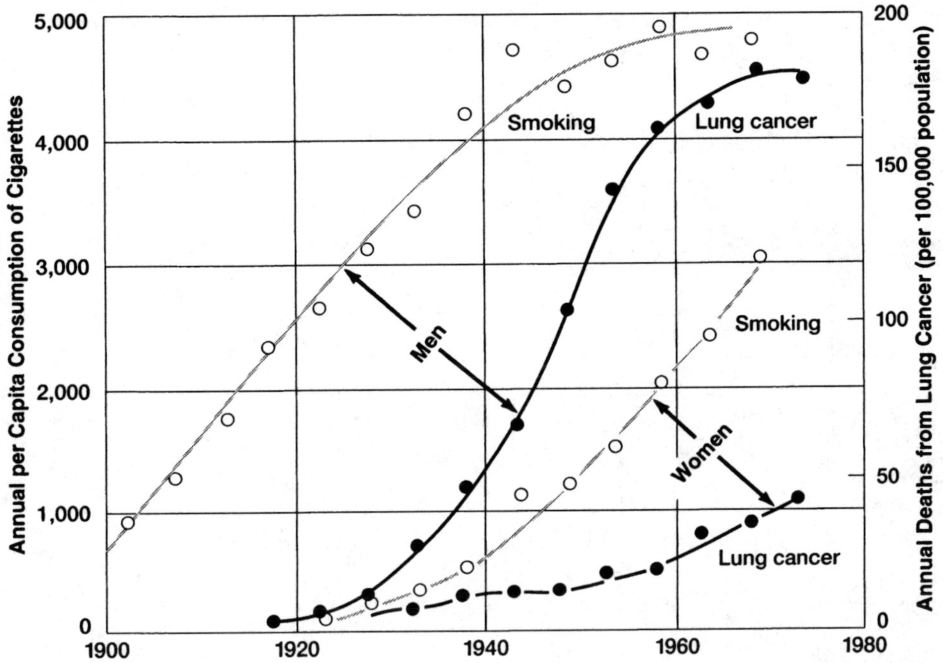

Why was lung cancer only examined in this study if cancer can arise in any part of the body? According to the graph in figure 17.1, the incidence of lung cancer does not show an increase for 10–20 years after smoking increases. How would you interpret this observation?

The results in figure 17.1 clearly indicate a relationship between an environmental factor (tobacco smoke) and lung cancer and support the hypothesis. This is, however, only one example. Are there other instances of environmental factors related to cancer incidence?

Extensive studies have been made of several human populations to investigate the incidence of cancer under different conditions. In one such study, the kinds and frequency of cancer were compared in native Japanese populations, Japanese immigrants to California, sons of Japanese immigrants, and California whites.

What do you predict would be found in this comparison?

Prediction 3—If environmental factors are causing cancer, then cancer rates should be characteristic of where the populations are living.

Result 3—Figure 17.2 shows the results of this comparison for 4 different kinds of cancer. Although there are significant differences between native Japanese and California whites, these differences are lost after 2 generations. This strongly supports the hypothesis that environmental factors determine the kind and incidence of cancer.

Figure 17.2 Cancer incidence in native Japanese, immigrants, and California whites.

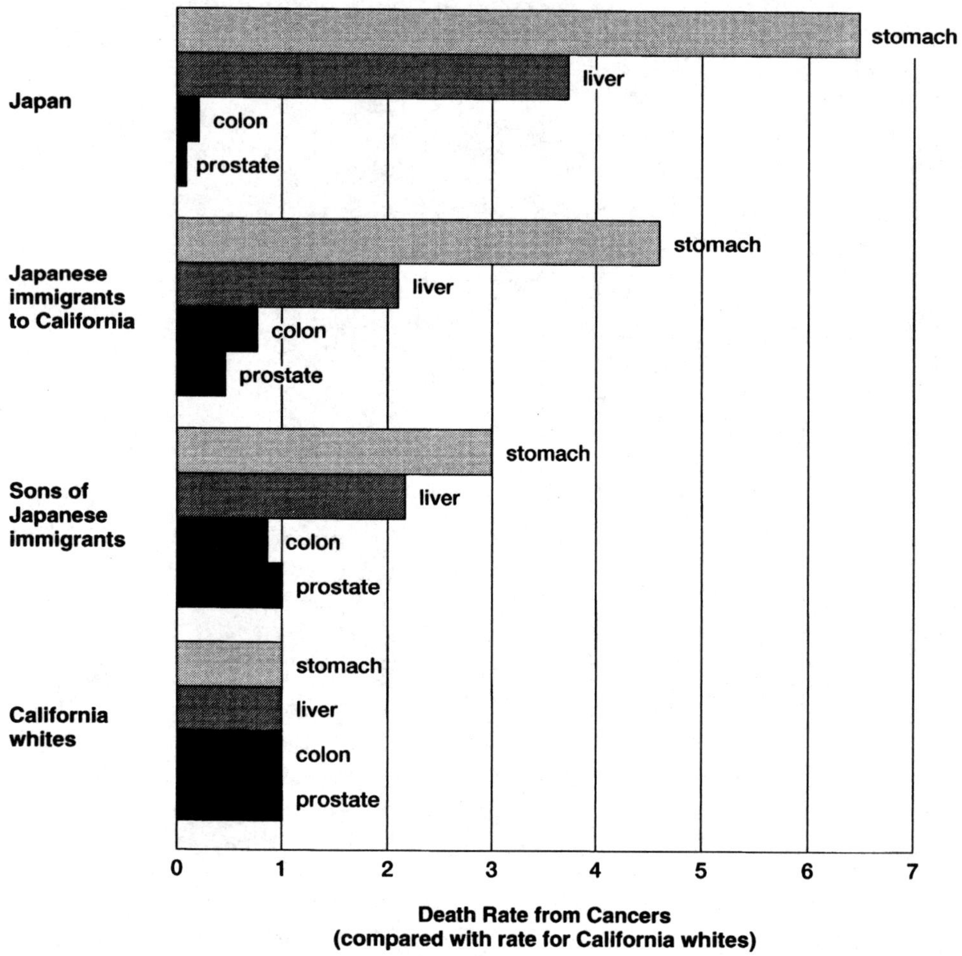

II Japanese immigrants and their offspring tend to marry within their ethnic group. How would this support the argument that environmental factors, rather than genetic factors, cause cancer? How would you design an experiment to investigate the hypothesis that exposure to sunlight causes skin cancer?

Although the data here clearly support the original hypothesis, there are some studies indicating that genetics is a factor. For example, breast cancer is twice as high in women whose mothers have had breast cancer. Such indications are in the minority, however, and environmental factors appear to be the most serious causative agent. If these environmental factors can be identified and exposure minimized, there should be a considerable decrease in cancer incidence.

Reference

Cairns, J. November 1975. The cancer problem. *Scientific American,* 64–78.

Case Study 17 THE CAUSES OF CANCER

Name ______________________
Section ______________________
Date ______________________

I

Question 1—Why was lung cancer only examined in this study if cancer can arise in any part of the body?

Answer 1

Question 2—According to the graph in figure 17.1, the incidence of lung cancer does not show an increase for 10–20 years after smoking increases. How would you interpret this observation?

Answer 2

II

Question 1—Japanese immigrants and their offspring tend to marry within their ethnic group. How would this support the argument that environmental factors, rather than genetic factors, cause cancer?

Answer 1

Question 2—How would you design an experiment to investigate the hypothesis that exposure to sunlight causes skin cancer?

Answer 2

18 Case Study: Human Locomotion (Difficulty Level 2)

Locomotion in walking and running expends a considerable amount of the total energy utilized by humans. For example, an average-sized person expends about 100 calories per hour standing upright but almost 240 calories per hour walking rapidly. When running, the energy expenditure increases to over 1,000 calories per hour. Such energy demands represent a major portion of the approximately 4,000 calories per day expended by the average person. Thus, it is logical to propose that movements in locomotion are adapted to make the most efficient and effective use of this energy expenditure.

Studies of the energy utilization in walking and running examine the body movements during locomotion and the effects these movements have on energy consumption. As a hypothesis, the following can be proposed:

Hypothesis—The mechanisms and movements of walking and running are adapted to make the most efficient use of energy.

Before testing this hypothesis, it is necessary to examine closely the body movements that take place in running and walking. As shown in figures 18.1 and 18.2, walking and running have characteristic gaits with some common features and other distinctly different features. Figure 18.1*d* shows that in walking *both* feet are always on the ground for part of the stride. Running, on the other hand, always has *both* feet off the ground for part of the stride (fig. 18.2).

Obviously, walking is at slower speeds and running at faster speeds; the faster the movement, the greater the energy consumption. Various methods could be used to measure energy consumption, but the simplest technique is to measure the amount of oxygen utilized during any particular activity such as walking and running at various speeds. As shown by equation 1,

Equation 1

$$\text{food}\begin{pmatrix}\text{protein}\\\text{carbohydrates}\\\text{fats}\end{pmatrix} + O_2 \rightarrow CO_2 + H_2O + \text{energy}$$

the utilization of O_2 releases energy in the body, which is used by muscles to do work. Thus, if the amount of O_2 used is known, the amount of energy generated can be determined. Careful measurements have shown that regardless of the food source (carbohydrates, proteins, or fats), about 4.8 calories are generated for every liter of O_2 consumed.

When O_2 utilization is measured at various speeds of walking and running, what do you predict would be observed?

Prediction 1—The energy used for walking will be less than the energy used when running. Moreover, the energy used to run at slower speeds will be greater than the energy for walking at these slower speeds. (If this were not true, we would expect people to run very slowly rather than walk to use energy efficiently.) The reverse should be true for running and walking at faster speeds.

Result 1—As shown in figure 18.3, walking requires a rapidly increasing energy expenditure up to a speed of about 2.5 m per second. In a similar manner, the energy requirements for running increase as the speed of running increases. Of particular interest is the point at about 2.5 m per second, where running and walking overlap—at this point, it is apparent that walking uses more energy than running. In fact, the energy required for walking is increasing very rapidly as the walking speed increases. Walking faster would seriously increase energy demands above that required for running. About 3 m per second is the maximum walking speed for most people; to move faster would require changing to a running stride. Therefore, walking appears to be very efficient up to speeds slightly above 2 m per second, but at higher speeds, running is the more efficient gait.

Figure 18.1 Successive stages in walking. At some stage in the stride, both feet are on the ground at the same time (*d*). The arrows indicate the force exerted by the feet on the ground. *(From American Scientist,* Walking and Running *by R. McNeill Alexander, Vol. 72, No. 4, pg. 349, July/August 1984.)*

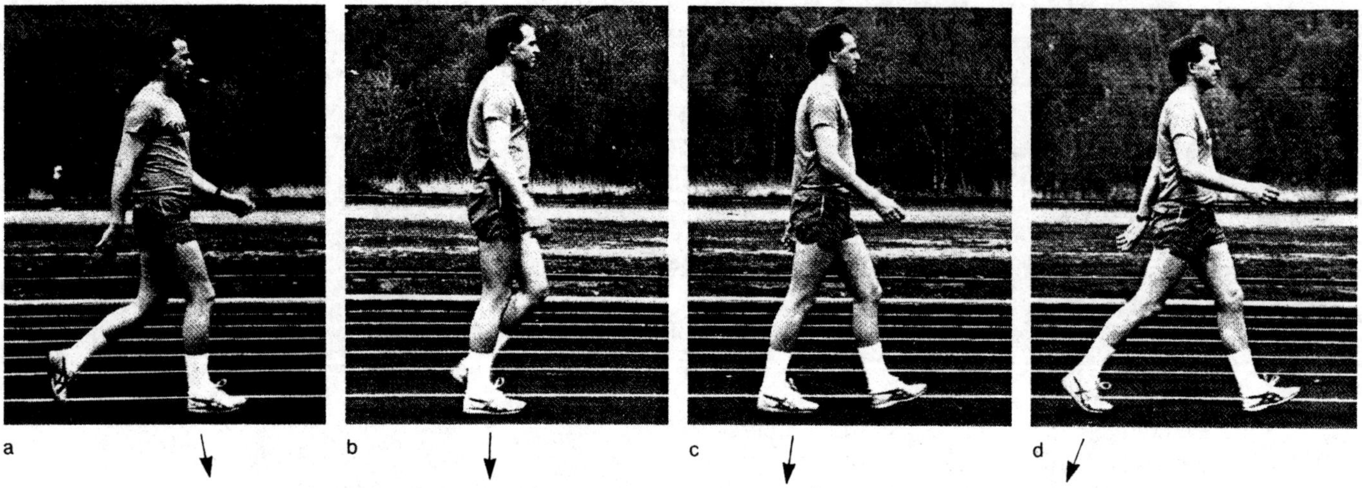

Figure 18.2 Successive stages in running. At some stage in the stride, both feet are off the ground at the same time. The arrows indicate the force exerted on the ground by the feet. *(From American Scientist,* Walking and Running *by R. McNeill Alexander, Vol. 72, No. 4, pg. 349, July/August 1984.)*

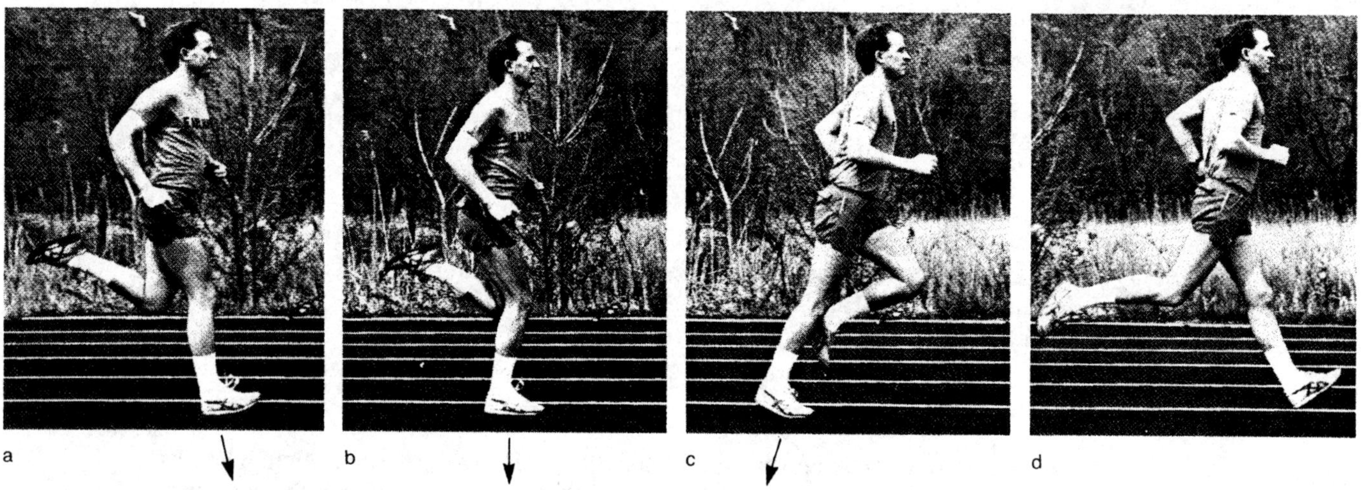

Figure 18.3 Energy expended in walking, running, and cycling at various speeds. Power is the number of calories expended per unit time and represents the rate of energy use.

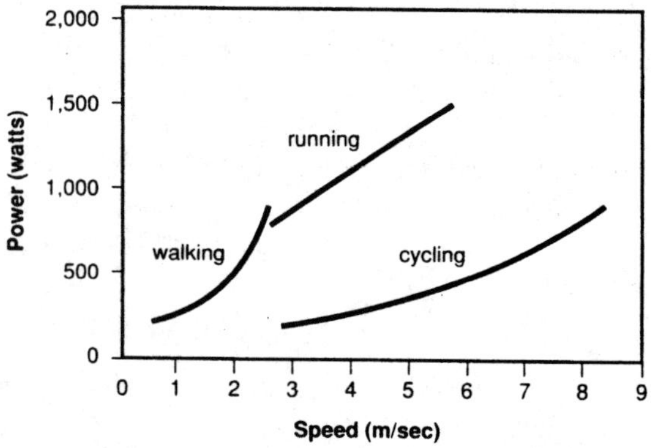

Question—Why is it logical that walking or running faster uses increasing amounts of energy?
Answer—As a person moves faster, the legs must move more rapidly. This means that the leg muscles are contracting at a faster rate and utilizing more energy.
Question—As a person walks more and more slowly, his or her energy requirement would approach what value?
Answer—Since standing still requires 100 calories per hour, energy requirements for slower and slower walking should approach 100 calories per hour.

I As a person walks faster, how would O_2 consumption change?
A change in what other gas could be measured to determine energy utilization?

Figure 18.3 also shows that the energy used in cycling is much lower than the energy required for either running or walking. Why should walking and running require so much more energy? A reexamination of figures 18.1 and 18.2 may provide a clue to answering this question. The arrows at the bottom of the figures indicate the force exerted when the foot strikes the ground. What would you predict about how this force might require extra energy in the mechanics of movement?
Prediction 2—If the force exerted by the feet on the ground results in an increase in energy requirements, this could account for the increased energy requirements of walking and running above that required for cycling.
Result 2—Whenever the front foot strikes the ground in either walking or running, there is some force exerted in the forward direction. This forward force will tend to slow down the speed of the walker or runner and is equivalent to putting on the brakes of a bicycle periodically and then speeding up again. Since running or walking requires that a foot be placed ahead of the body when moving forward, a forward force will be exerted while having a braking effect on the movement, which cannot be avoided.

II In which parts of figures 18.1 and 18.2 is a force exerted to move the person forward?
If a runner "overstrides" (placing a foot farther forward than in a normal stride), will this have a greater or lesser braking effect?

A detailed analysis of the "braking" force and the energy that could be lost in the braking effect in running has indicated that running should require *much more* energy than indicated in figure 18.3. Slow running should require 1.8 times as much energy as shown, while faster running should utilize 3 times as much energy. Why is running so much more efficient than the analysis indicates?

The structure of the leg provides a hint to how energy could be conserved. When the foot strikes the ground in running (fig. 18.2*a*), the foot and leg absorb the braking energy. If this energy could be stored and retrieved when the leg provides the forward thrust (fig. 18.2*c*), instead of being lost in the braking action, energy would be used much more efficiently. One way to store the braking energy is in the stretch of the leg tendons. The tendons could act as a spring that is stretched in figure 18.2*a* and recoiled in figure 18.2*c*, providing extra push and requiring less energy from the muscle in the forward thrust. Is such energy storage possible in the tendons?

Prediction 3—If the leg tendons could store and release energy as a spring would do, energy would be more efficiently utilized.
Result 3—Tests with tendons have shown that when stretched, as much as 93% of the stored energy can be returned when the tendon recoils. Moreover, the Achilles tendon connecting the calf muscles to the heel is located in a position to stretch and return energy when running. Thus, the stretch and recoil of tendons is an effective method for making efficient use of energy.

III What experimental result led to the search for an explanation of the higher than expected energy efficiency in running?
How would running (a) uphill and (b) downhill affect the energy use curves in figure 18.3?
The energy consumption curve for walking in figure 18.3 is not a straight line, indicating that walking becomes increasingly less efficient as the speed increases. Why might this be so?

Although the mechanism of walking and running involves some inevitable energy loss (i.e., in the braking effect), both the gait (i.e., walking or running) and energy conservation (i.e., tendon stretch) have been shown to be adapted to efficient energy utilization. The hypothesis receives strong support from the results, revealing how energy is used in human locomotion.

References

Alexander, R. M. July–August 1984. Walking and running. *American Scientist,* 348–54.
Davidovits, P. 1975. *Physics in Biology and Medicine.* Englewood Cliffs, N.J.: Prentice-Hall.

Name ____________________
Section ____________________
Date ____________________

I

Question 1—As a person walks faster, how would O_2 consumption change?

Answer 1

Question 2—A change in what other gas could be measured to determine energy utilization?

Answer 2

II

Question 1—In which parts of figures 18.1 and 18.2 is a force exerted to move the person forward?

Answer 1

Question 2—If a runner "overstrides" (placing a foot farther forward than in a normal stride), will this have a greater or lesser braking effect?

Answer 2

III

Question 1—What experimental result led to the search for an explanation of the higher than expected energy efficiency in running?

Answer 1

Question 2—How would running (a) uphill and (b) downhill affect the energy use curves in figure 18.3?

Answer 2

Question 3—The energy consumption curve for walking in figure 18.3 is not a straight line, indicating that walking becomes increasingly less efficient as the speed increases. Why might this be so?

Answer 3

19 Case Study: Hibernation (Difficulty Level 2)

During winter months, when the productivity of the environment is minimal, animals can face severe difficulties obtaining sufficient food. This is particularly true for warm-blooded animals such as mammals, where more than half of the food intake is used in generating heat to maintain higher body temperatures. Moreover, these energy demands are even greater in winter, when more energy must be expended to keep the body temperature constant.

Maintaining a higher and constant body temperature (termed *euthermy*) is considered a major survival advantage of mammals. The disadvantage, as mentioned, is that euthermy requires a higher food intake to maintain the higher body temperature during the winter months, when sufficient food is not available.

Many animals are able to offset this disadvantage by decreasing their metabolic rates, and thus energy utilization, by hibernating during the winter. Even during hibernation, however, *all* mammals periodically arouse and rewarm for short periods without leaving their dens. Apparently, euthermy provides such a decided adaptive advantage that periodic rewarming is advantageous even when winter months require decreased metabolism.

Mammals seem to be responding to 2 conflicting needs: the need to maintain higher body temperature while at the same time the need to conserve energy. Hibernation, and the periodic rewarming during hibernation, reflect the response to these conflicting needs.

Two other factors should also be considered when examining the hibernation patterns of mammals. First, larger animals can store more fat (which will provide energy during the winter months) and at the same time expend a smaller proportion of their body fat to maintain a higher body temperature than can small animals. This is because smaller animals lose body heat more rapidly to the environment than do larger animals. Therefore, small animals are at a double disadvantage; they store less fat and lose body heat at a more rapid rate.

Second, larger animals require a longer reproductive period because the gestation period (the time from conception to birth of the young) and the time needed to care for the young is longer. The amount of time spent in hibernation might be limited for larger animals because of this longer time required for gestation and infant care. Smaller animals, on the other hand, exhibit relatively short reproductive periods.

Based on a consideration of these 2 factors, the following hypothesis can be proposed:

Hypothesis—Large and small mammals will exhibit differences in hibernation patterns that allow the most effective use of energy and maximize reproductive success.

To test this hypothesis, several measurements and observations can be made. The first factor considered previously indicated that larger animals can store and make more efficient use of body fat (i.e., energy reserves). When the length of hibernation is measured in large and small animals, what would you predict would be observed?

Prediction 1—Since larger animals can store relatively more body fat and use this energy source more efficiently (i.e., by losing a smaller proportion of their body heat to the environment), they should hibernate for longer periods of time.

Result 1—Contrary to the prediction, larger animals are *not* observed to hibernate longer than smaller animals. Detailed studies have indicated, however, that rewarming of larger animals during hibernation is for longer periods as shown in figure 19.1. This result is consistent with the argument that higher body temperature is a selective advantage and supports the hypothesis by indicating one difference in hibernation patterns of different species.

Question—If larger animals *did* hibernate for a longer period, what problem might occur?

Answer—Since larger animals require a longer reproductive period for gestation and care of the young, more time spent in hibernation would be a serious disadvantage in terms of reproduction.

Would mammals be expected to hibernate in warmer climates?
Would a large animal or a small animal be more likely to use up a greater proportion of body fat during hibernation?

Figure 19.1 Relationship between the size of an animal and the duration of the periodic rewarming periods (euthermic intervals) during hibernation.

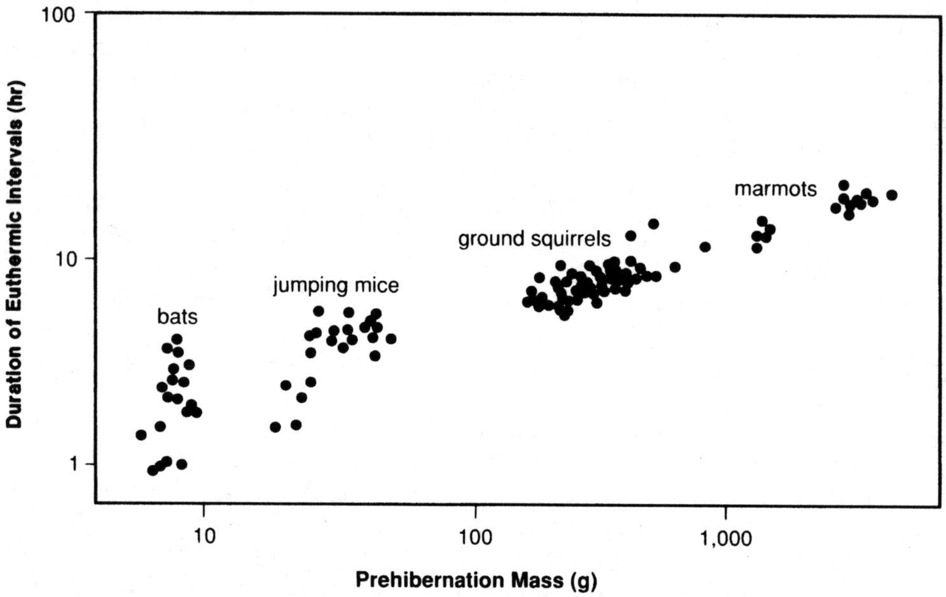

The second factor to consider, the need for a sufficiently long reproductive period, also suggests differences that might occur between hibernating species. Since larger animals require a longer period for reproduction in the spring and summer, what would you predict would be observed when comparing the hibernation patterns of large and small mammals?

Prediction 2—Larger animals should emerge from hibernation and breed earlier than smaller animals. In this way, a longer time period will be available to the larger animals for gestation and caring for young animals before the next winter season.

Result 2—Investigation and comparison of the hibernation patterns of marmots, ground squirrels, and jumping mice yielded the following results. These animals live in the same environment in mountainous areas of the western United States, where the winters are usually harsh.

Marmots were the largest of the animals investigated. Hibernation is terminated spontaneously, and adults emerge and breed well before food is available in the spring. Because they are larger animals, marmots are able to continue to live on stored fat in the spring until food is available.

Ground squirrels are 10 times smaller than marmots and exhibit more frequent rewarming periods in the spring. During these periods, the females apparently observe the environment and emerge when food is available. Males, however, terminate hibernation spontaneously and emerge earlier than females.

Jumping mice are 10 times smaller than ground squirrels. These animals also rewarm more frequently in the spring but do not terminate hibernation until the snow has melted, the ground has warmed, and ample food is available.

These comparisons indicate that larger animals break hibernation earlier in the spring. Marmots, due to their larger size and greater energy storage, can afford to "gamble" on an early emergence from hibernation and in fact must do so to insure sufficient time for reproduction. The much smaller jumping mice can be more careful and conservative, waiting until later in the spring to emerge, and they require far less time than marmots for reproduction.

Question—How would the hibernation patterns of even larger hibernating animals, such as bears, be adapted to the needs of reproduction?

Answer—In bears, gestation and care of the young require even longer than the time available in the spring and summer. As a consequence, bears breed *before* hibernation. Birth and breast feeding occur during hibernation, and young grow and become relatively independent during the subsequent warm seasons.

II What advantage is there to male ground squirrels emerging from hibernation before females? In contrast to ground squirrels, *both* male and female jumping mice emerge from hibernation after warming of the environment at the same time. Why would this be so?

Result 2 indicates that reproductive needs apparently have a strong effect on the evolution of hibernation patterns. This effect may also be apparent in other comparisons. In another test, the hibernation patterns of adult male marmots were compared with those of juvenile male marmots who had not yet reached breeding age.

What differences would you predict in the patterns?

Prediction 3—Adult males should terminate hibernation earlier than nonbreeding juveniles. The juveniles would have no reproductive advantage by early termination since they do not breed but *would* increase their chances of survival by extending hibernation.

Result 3—As shown in figure 19.2, juvenile male marmots hibernate for much longer periods than do adult male marmots. This result indicates an obvious difference in hibernation patterns within the same species.

III Would you predict a similar difference in hibernation patterns between adult and nonbreeding juvenile *female* marmots?

Which species (marmot, ground squirrel, or jumping mouse) would be more seriously affected by a long cold spring in which the snow remains on the ground for a longer period?

These results show that hibernation patterns of animals indicate the various strategies that accommodate the animals' need to make the most effective use of energy stores and fulfill reproductive requirements. Various strategies provide obvious benefits to animals of different size and provide strong support for the hypothesis that such different patterns will be observed.

Figure 19.2 Comparison of hibernation patterns in adult male and nonbreeding juvenile male. Notice the juvenile exhibits about the same duration of dormant periods without rewarming but hibernates about two months longer.

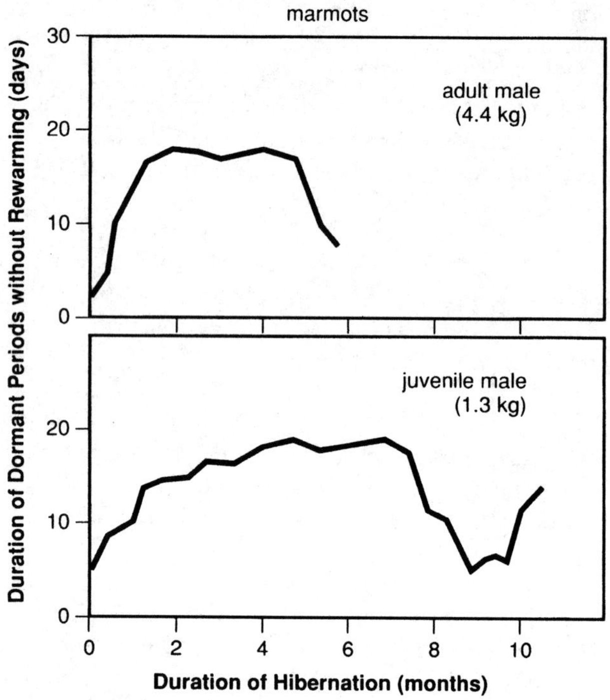

Reference

French, A. R. November–December 1988. The patterns of mammalian hibernation. *American Scientist,* 568–75.

Case Study 19 HIBERNATION

Name ____________________
Section ____________________
Date ____________________

I

Question 1—Would mammals be expected to hibernate in warmer climates?

Answer 1

Question 2—Would a large animal or a small animal be more likely to use up a greater proportion of body fat during hibernation?

Answer 2

II

Question 1—What advantage is there to male ground squirrels emerging from hibernation before females?

Answer 1

Question 2—In contrast to ground squirrels, *both* male and female jumping mice emerge from hibernation after warming of the environment at the same time. Why would this be so?

Answer 2

III

Question 1—Would you predict a similar difference in hibernation patterns between adult and nonbreeding juvenile *female* marmots?

Answer 1

Question 2—Which species (marmot, ground squirrel, or jumping mouse) would be more seriously affected by a long cold spring in which the snow remains on the ground for a longer period?

Answer 2

20 Case Study: Boundary Layers (Difficulty Level 3)

Boundary layers arise in a fluid (either liquid or gas) near a stationary surface and have been proposed as a factor that can significantly affect physiological processes. A boundary layer is defined as a layer of relatively unmoving fluid and is formed due to friction between the stationary surface and the surrounding moving fluid as illustrated in figure 20.1. As the speed of the flowing fluid increases, the thickness of the boundary layer is decreased; friction between the fluid more distant from the stationary surface and the boundary layer will tend to "drag" along fluid in the boundary layer and thus decrease boundary layer thickness as the speed of fluid flow increases.

The potential physiological effects of boundary layers can be visualized by considering the flow of water over a gas exchange organ such as the gills in a fish. If the stationary surface in figure 20.1 is the external surface of a gill plate and the boundary layer is the layer of water next to the gill surface, gas exchange with external water will first occur in the boundary layer. Carbon dioxide from the gill capillaries released to the external environment will enter the boundary layer and will tend to accumulate since this water layer is not well mixed with the surrounding fluid. This increasing CO_2 concentration will decrease the concentration gradient and thus decrease the net movement of CO_2 from the blood to external water. A similar effect would be expected for O_2 exchange, where O_2 concentration would be decreased in the boundary layer.

Such reasoning has led to the following hypothesis:

Hypothesis—Boundary layers can exert a significant effect on physiological processes that depend on gas exchange with the external environment.

How could this hypothesis be tested?

In one experiment by Westlake (1966), the rate of photosynthesis in aquatic plants was measured as a function of the velocity of water flow across the plants. The plants were maintained in a transparent closed container with openings at each end. Water could be circulated through the chamber by a pump and the speed of the pump varied to control the rate of water flow. Photosynthetic rate was measured by the difference in O_2 content of the water entering and leaving the chamber.

Prediction 1—If boundary layer thickness affects gas exchange, then the rate of photosynthesis should increase as the rate of water flow increases and the boundary layer becomes thinner.

Result 1—An increase in photosynthetic rate is observed as water flow velocity increases as shown in figure 20.2. This result is consistent with the hypothesis. Apparently the boundary layer thickness can limit the rate of gas exchanges and thus photosynthesis.

Figure 20.1 The arrows representing the velocity of the moving fluid. The fluid immediately adjacent to the stationary surface will *not* be moving.

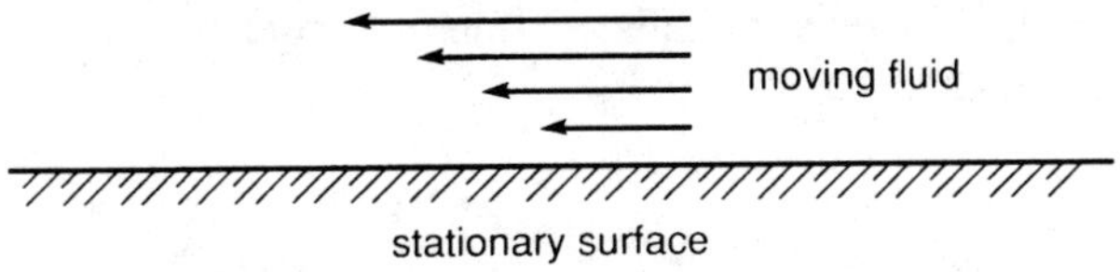

Figure 20.2 Rate of photosynthesis (oxygen production) by an aquatic plant as the velocity of water flow over the surface of the plant is varied.

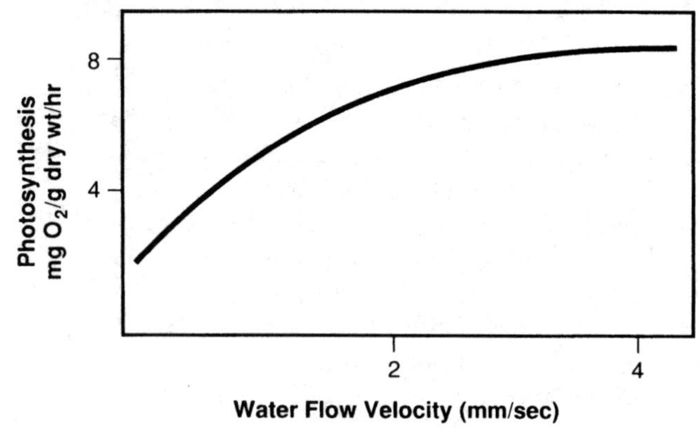

I Would O_2 be higher or lower in the boundary layer, relative to the moving fluid? Why? Could changes in the concentration of another gas have been measured? If so, which one? The curve in figure 20.2 appears to be leveling off (i.e., reaching a stable rate that will not increase further with increasing water velocity). How would you explain this observation?

In another experiment, Gerard and Mann (1979) studied large marine algae (kelp) under natural conditions at 2 sites. At one site in the open ocean, wave action created rapid water movement around the plants. At the other site, plants were protected from wave action, and water movement was considerably less than in the open ocean. Growth rate, measured as the rate of leaf elongation, was measured in plants at the 2 sites.

If boundary layers affect gas exchange, then what would you predict about the growth rates of plants at the 2 sites?

Prediction 2—Growth rates of plants in the open ocean should be higher since the more rapid rate of water movement would decrease the boundary layer thickness, increase gas exchange, and thus increase the rate of photosynthesis.

Result 2—As predicted, higher growth rates were observed in plants from the open ocean.

In a subsequent experiment, plants were taken from the open ocean and transplanted at the protected site. In this measurement, growth rates were compared between open ocean and transplanted plants.

What would you predict in this case?

Prediction 3—Growth rate should be greater in plants growing in the open ocean.

Result 3—As shown in figure 20.3, during most of the year, open ocean plants exhibited significantly higher growth as predicted. Again, these results support the hypothesis.

II Why was this experiment performed? What information does this experiment provide that could not be drawn from Result 2?

How would you explain the observation that the growth rates were the same in the spring of 1978?

Although the previous experiment yielded results that supported the hypothesis, the organisms under investigation are highly complex systems with numerous variables. These variables could be affecting photosynthesis and growth in ways independent of any proposed boundary layer. A more simple system where the variables could be more carefully controlled was utilized in an experiment by Hays and Franki (1970). In their investigation, the multicellular membrane of a toad bladder was used to separate 2 chambers filled with water. The passage of water through the membrane could be monitored by adding radioactive water to one compartment and measuring how quickly the radioactivity appeared in the second compartment. The water in the compartments could be stirred at various speeds and thus increase the speed of movement of water over the surface of the membrane. In this way, the thickness of the boundary layer could be decreased at higher stirring speeds.

Figure 20.3 Growth rate in plants at exposed and sheltered sites.

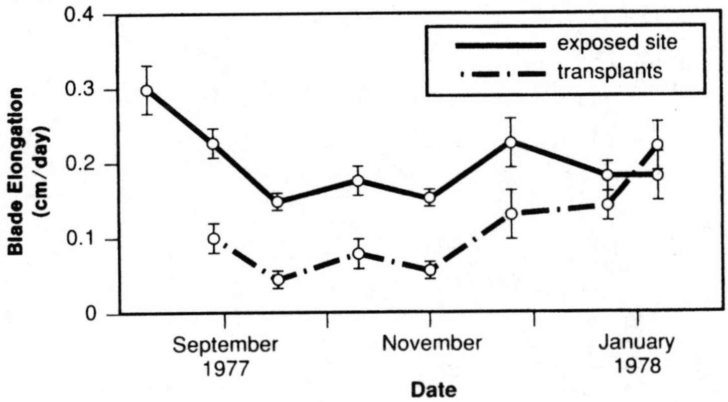

This experiment was performed in 2 ways: first with an untreated membrane (control) and second with a membrane treated with vasopressin, a hormone that increases the permeability of the toad bladder membrane. Permeability is a measure of the ease with which a substance can cross a membrane; the higher the permeability of a membrane to particular material, the more easily it can cross the membrane.

Prediction 4—If boundary layers affect the rate of movement of material across the membrane, then movement (i.e., the rate at which radioactive water appears in the second chamber) should increase as the speed of stirring increases. Moreover, a greater change in rate of movement should be observed when the permeability of the membrane has been increased by treatment with vasopressin.

Result 4—The results are shown in figure 20.4. As predicted, increased stirring speed increases the rate of movement across the membrane, indicating that decreasing the thickness of the boundary layer increases movement of material across the membrane.

This experiment also indicates that when the membrane permeability is high (i.e., material moves more easily across the membrane), the boundary layer or layers exert an increasingly significant effect on movement. In other words, a major part of the resistance to movement is due to the boundary layer. When the boundary layer is removed or its thickness decreased, movement across the membrane increases dramatically as in the vasopressin-treated membrane. On the other hand, the untreated membrane has a lower permeability and is the major barrier to movement; removing the boundary layers has relatively little effect.

III The permeability is almost the same for the untreated membrane and the vasopressin-treated membrane when the stirring speed is low. Why would this be so?

In this experimental arrangement, the water on *both* sides of the membrane is stirred. What effect would be observed if only one chamber was stirred?

The membranes where primary gas exchange occurs in organisms (i.e., gills, lungs, stomatal apertures) are necessarily thin and highly permeable. As shown in these experiments, boundary layers can significantly reduce movement across permeable membranes and, thus, considerable support is provided for the hypothesis that boundary layers exert a significant effect on physiological processes.

Figure 20.4 Permeability of toad bladder to water as stirring speed is varied.

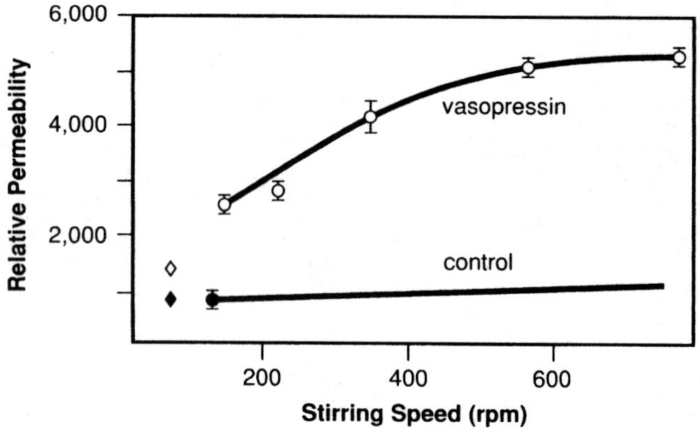

References

Gerard, V. A., and Mann, K. H. 1979. Growth and production of *Laminaria lonquicruris* (Phaeophyta) populations exposed to different intensities of water movement. *J. Phycol.* 15:33–41.

Hays, R. M., and Franki, N. 1970. The role of water diffusion in the action of vasopressin. *J. Membrane Biol.* 2:263–76.

Westlake, D. F. 1966. Some effects of low-velocity currents of the metabolism of aquatic macrophytes. *J. Exptl. Botany* 18:187–205.

Case Study 20 BOUNDARY LAYERS

Name ______________
Section ______________
Date ______________

I

Question 1—Would O_2 be higher or lower in the boundary layer, relative to the moving fluid? Why?

Answer 1

Question 2—Could changes in the concentration of another gas have been measured? If so, which one?

Answer 2

Question 3—The curve in figure 20.2 appears to be leveling off (i.e., reaching a stable rate that will not increase further with increasing water velocity). How would you explain this observation?

Answer 3

II

Question 1—Why was this experiment performed? What information does this experiment provide that could not be drawn from Result 2?

Answer 1

Question 2—How would you explain the observation that the growth rates were the same in the spring of 1978?

Answer 2

III

Question 1—The permeability is almost the same for the untreated membrane and the vasopressin-treated membrane when the stirring speed is low. Why would this be so?

Answer 1

Question 2—In this experimental arrangement, the water on *both* sides of the membrane is stirred. What effect would be observed if only one chamber was stirred?

Answer 2

21 Case Study: Aneurysms (Difficulty Level 3)

Although arteries are the strongest of the body's blood vessels, they are not rigid tubes for conducting blood. The arterial wall is thick, as shown in figure 21.1, but it is also extensible and stretches when blood pressure increases during contraction of the heart (systole). When the blood pressure decreases (diastole), the arterial wall contracts, much as a stretched rubber band contracts when a force is released. Thus, the artery expands and contracts like an elastic tube in response to the pressure change during each heartbeat.

Under this repeated stress of stretching, it is not surprising that parts of an artery can become permanently stretched into a balloonlike structure termed an *aneurysm.* If undetected, an aneurysm can rupture, and this often leads to death.

The most common site of aneurysms in humans is in larger arteries, particularly at the lower end of the abdominal aorta, where the artery branches into 2 smaller arteries. Why should this position on the aorta be especially prone to the development of an aneurysm? Are there particular structural features at this site that can lead to an aneurysm?

These observations and questions have led to the following hypothesis:

Hypothesis—Specific properties of the lower aorta favor aneurysm formation.

Testing of this hypothesis has involved investigations of the structural and physiological characteristics of the aorta and of developing aneurysms.

As a first step in testing the hypothesis, a more detailed examination and analysis of the lower aorta was made. This analysis considered the physical characteristics of a site where branching of a large artery forms 2 smaller arteries. The alternating pressure exerted by the heart on the blood travels through the aorta as a pulse or wave of pressure. At a branch point, a pressure pulse is reflected back to some degree and could place greater stress on the aorta wall. Is there reason to believe that this reflection could play a role in aneurysm formation?

Prediction 1—An analysis of the properties affecting the strength and nature of the reflected pressure wave should indicate potential variables that could lead to an aneurysm.

Result 1—A mathematical analysis of the variables affecting reflection of the pressure wave indicate that the strength of the reflected wave depends on the relative cross-sectional areas of the aorta and branch arteries (fig. 21.2). Moreover, as a person ages, the aorta relaxes to some extent, and the cross-sectional area increases. This increases the strength of the reflected wave and thus increases the stress on the wall of the aorta. The structure of the branching aorta therefore produces reflected pressure waves that produce stress in the artery wall.

Figure 21.1 Comparison of wall structure in arteries, veins, and capillaries. Notice the much thicker wall in arteries.

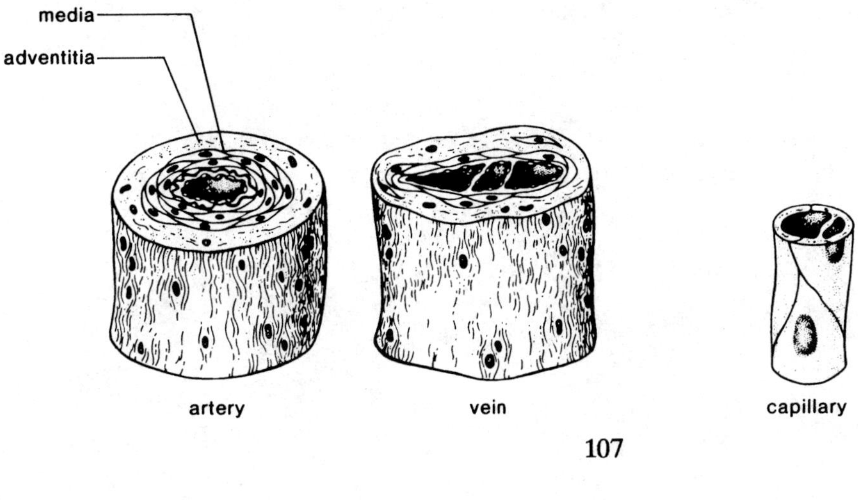

Figure 21.2 Structure of the lower abdominal aorta. The strength of the pressure wave reflected from the branch point is proportional to b/2a, the relative diameters of the aorta and the branch arteries. The diagram also shows the beginning of an expansion to form an aneurysm. Notice that blood flow becomes turbulent at the expansion.

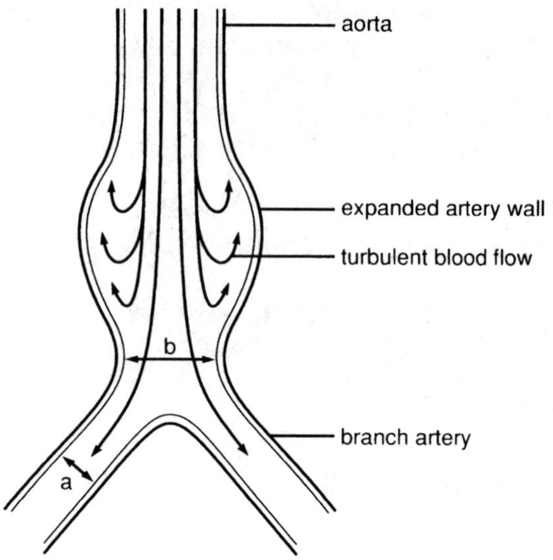

I Why is pressure wave reflection especially important in the aorta?
Would it be as important in a large vein? Why or why not? (Hint: Consider the pulse of blood flow in an artery compared to a vein.)

Turbulence is another aspect of blood flow that has been suggested as a factor affecting aneurysm formation. When the aortic wall becomes slightly expanded, blood flow tends to become turbulent rather than smooth. This is because blood flowing near the wall moves more slowly, while blood in the center of the artery moves more rapidly (see Case Study 20, "Boundary Layers," for a discussion of friction between moving fluids and stationary surfaces). This effect is illustrated in figure 20.2.

Prediction 2—Turbulence in the flowing blood could create additional stress in the artery wall.

Result 2—Since turbulence is irregular flow, vibration can be created in the artery. Investigations have shown that low-frequency vibrations can cause destruction and deterioration in the artery wall. Such effects clearly can weaken the wall and enhance aneurysm formation. When subjected to the repeated pulsing vibration, the artery wall is stressed.

II Why would turbulence be a problem only after artery wall expansion has started?
Could turbulence be a problem at other points in the circulatory system? Why or why not?

Detailed investigations have been conducted of the tissue structure of the aortic wall. Examinations of the media, the middle layer of the artery wall, which is responsible for most of the wall strength, have indicated that the human abdominal aorta may be unexpectedly weak. Comparisons to experimental animals lead one to expect 35 layers or lamellae in the media of the human abdominal aorta. In actuality, only 28 are found.

Prediction 3—The smaller number of lamellae should result in a weaker aortic wall.

Result 3—Measurements and calculations have shown that the tension per lamella in the abdominal aorta of humans is more than double the tension in lamellae of experimental animals. It has also been found that the tension per lamella in the human thoracic aorta is about one-half that found in the abdominal aorta and that aneurysms in the thoracic aorta are rare.

III Wouldn't you expect that evolution and natural selection would have resulted in a stronger abdominal aorta? Why do you think this has not occurred?

A further analysis of the artery wall indicates another serious complication with aneurysms. In an expanding aneurysm, it has been argued that the tension in the wall becomes greater when the aneurysm expands. This is based on the calculation that shows that wall tension (T) is proportional to $Pr/_t$, where P is the blood pressure inside the aneurysm, r is the radius of the aneurysm, and t is the artery wall thickness.

Does this analysis predict a weakened aneurysm wall?

Prediction 4—If tension (T) is proportional to $Pr/_t$, then as the radius increases, tension will increase, and as the aneurysm expands, t will become thinner, further increasing the tension. If this expression is valid, it would be expected that because the wall is thinner and the radius greater, the wall tension will become greater as the aneurysm expands.

Result 4—Calculations show that the wall tension in an aneurysm with a radius of 3 cm is 12 times the tension in the normal aorta.

The aneurysm, then, creates a situation where once the aneurysm starts to develop, the wall tension increases, and this results in further aneurysm expansion and increased wall tension. This continues until the aneurysm ultimately ruptures.

IV What would be the effect of blood pressure on wall tension?

This analysis strongly suggests that several factors favor aneurysm formation at the lower abdominal aorta. Considerations of blood flow, wall structure, and circulatory system structure all contributed to the analysis. The investigation of aneurysms is a good example of the fact that several different factors and approaches must be considered to investigate a question fully.

Reference

Johansen, K. July 1982. Aneurysms. *Scientific American,* 110–25.

Case Study 21 ANEURYSMS

Name ____________________
Section ____________________
Date ____________________

I

Question 1—Why is pressure wave reflection especially important in the aorta?

Answer 1

Question 2—Would it be as important in a large vein? Why or why not?

Answer 2

II

Question 1—Why would turbulence be a problem only after artery wall expansion has started?

Answer 1

Question 2—Could turbulence be a problem at other points in the circulatory system? Why or why not?

Answer 2

III

Question 1—Wouldn't you expect that evolution and natural selection would have resulted in a stronger abdominal aorta? Why do you think this has not occurred?

Answer 1

IV

Question 1—What would be the effect of blood pressure on wall tension?

Answer 1

22 Case Study: The Opening and Closing of the Stomata (Difficulty Level 3)

The guard cells on either side of a stoma control whether a stoma is open or closed. When the guard cells swell with water and develop turgor pressure, a stoma is open (fig. 22.1*a*), and when the guard cells lose water, a stoma is closed. Usually stomata are open during the day and are closed during the night. When stomata are open, water is transported in xylem to the leaves and the leaves receive CO_2 by way of the stomata. Water and CO_2 are the reactants needed for photosynthesis to occur.

Is it adaptive for plants to have their stomata open during the day? Why? Will the level of CO_2 inside the leaves be high or low when a plant is photosynthesizing? Why?

To test if both light and low level of CO_2 are needed to cause stomata to open, researchers exposed epidermal strips (the outer layer of leaf tissue containing epidermal cells and guard cells) floating on distilled water to CO_2-free and CO_2-laden air in both the light and in the dark. Should CO_2-free air in the presence of light result in a larger stomatal opening?

Prediction 1—The CO_2-free air in the presence of light should result in a larger stomatal opening.

Result 1—As shown in figure 22.1*b*, exposure of epidermal strips to CO_2-free air in the light did result in larger stomatal opening.

I

Which factor seems more pertinent to stomatal opening, concentration of CO_2 or presence of light?

In the dark a plant carries on cellular respiration and what gas is released? Now explain why a stoma provided with CO_2-free air in the dark initially starts to open and then retreats to partly closed.

Since a stoma never opens or closes completely in the dark, is it possible that some additional factor (besides CO_2 concentration) not tested for is involved in controlling stomatal movement?

To test whether inhibition of photosynthesis would have an effect on stomatal opening, inhibitors of photosynthesis were applied to epidermal strips exposed to light and CO_2-free air. In the light, a plant carries on both photosynthesis and cellular respiration; therefore, if photosynthesis is inhibited, the plant will only be respiring. Would you predict, then, that if photosynthesis is inhibited, a stoma will close?

Prediction 2—Inhibition of photosynthesis in the light should lead to stomatal closing.

Result 2—As predicted, inhibition of photosynthesis in the light did lead to stomatal closing.

II

What is a possible role of photosynthesis in causing the stomata to open? Is this a direct role or an indirect role?

Is it possible that photosynthesis has another effect not tested for?

This sequence of events has been suggested to explain stomatal opening: (1) low CO_2 concentration in the leaf in the presence of light leads to (2) high concentration of K^+ in guard cells, which leads to (3) uptake of water by guard cells, which leads to (4) opening of a stoma. If this sequence of events is to be expected, which of the experimental conditions noted in the graph should cause the greatest K^+ concentration in guard cells?

Prediction 3—The CO_2-free air in the presence of light should result in the greatest K^+ concentration in guard cells.

Result 3—Of the experimental conditions noted in figure 22.1*b*, CO_2-free air in the presence of light did result in the greatest K^+ concentration in guard cells. Indeed, for all conditions noted in the graph, measurements of the final K^+ concentrations did agree with what would be predicted by the graph.

III Explain why you would expect K^+ to increase initially and then decrease in guard cells of epidermal strips exposed to CO_2-free air in the dark.
Give a sequence of events, corresponding to events 1-4, that would explain why stomata, which are open during the day, close at night.

A logical question to ask is, What is the source of the K^+ that enters the guard cells when stomata open? The epidermal strips used in these experiments were floated on distilled water that lacked K^+; therefore, the only source is from the epidermal cells surrounding the guard cells.

Figure 22.1 *a.* This electron micrograph shows an open stoma and its guard cells. A stoma can vary as to the degree it is open. *b.* Epidermal strips (containing epidermal and guard cells only) floating in water were exposed to the 4 experimental conditions noted in the key. The experimenter recorded the degree of stomatal opening as related to time. *(©David M. Phillips/Visuals Unlimited)*

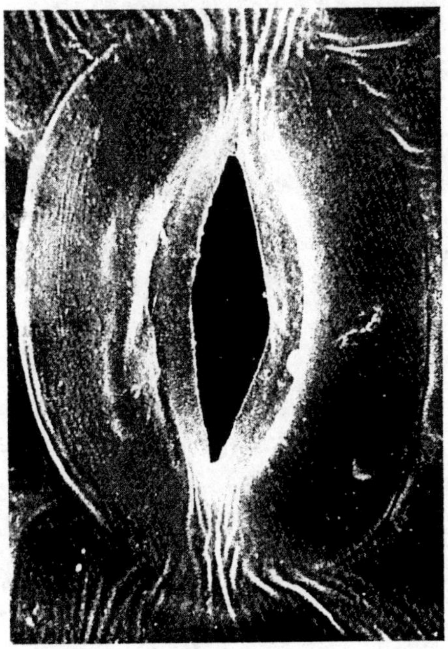

a.

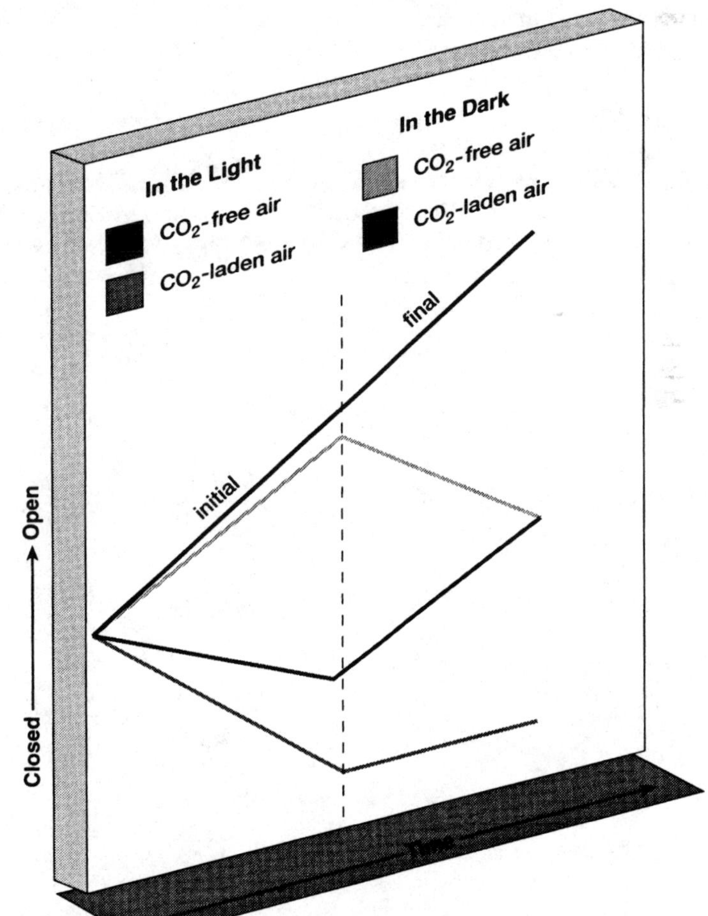

b.

If these epidermal cells supply K^+ to the guard cells, what do you predict would happen to the level of K^+ in epidermal cells as the concentration in guard cells increases and as the stomata open?
Prediction 4—K^+ concentration in epidermal cells should decrease as the concentration in guard cells increases and as the stomata open.
Result 4—Experimental measurements show that there was a decrease in epidermal cells K^+ concentration as the concentration in guard cells increased and as the stomata opened.

IV

If there was one epidermal cell per guard cell, what might happen to this epidermal cell once it gives up its K^+? Why?
Recognizing that there are many epidermal cells per guard cell, explain why you would not expect epidermal cells to suffer dehydration due to the passage of K^+ from epidermal cells to the guard cells.

The experiments described here support the hypothesis that in the light a low CO_2 concentration causes K^+ movement into guard cells, leading to increased turgor pressure and opening of the stomata. Other questions are raised, however, by these investigations. For example, what is the mechanism whereby CO_2 controls K^+ movement in the light, and what are the factors that control K^+ movement in the dark? It is customary for experiments to generate new questions and lead to further hypotheses for investigation.

References

Pallaghy, C. K. 1971. Stomatal movement and potassium transport in epidermal strips of *Zea mays:* The effect of CO_2. *Planta* (Berl.) 101,287–95, published by Springer-Verlag.
Rashke, K. 1975. Stomatal action. *Annual Revue Plant Physiology* 26:309–40.

Case Study 22 THE OPENING AND CLOSING OF THE STOMATA

Name ____________________
Section ____________________
Date ____________________

I

Question 1—Which factor seems more pertinent to stomatal opening, concentration of CO_2 *or* presence of light?

Answer 1

Question 2—In the dark a plant carries on cellular respiration and what gas is released? Now explain why a stoma provided with CO_2-free air in the dark initially starts to open and then retreats to partly closed.

Answer 2

Question 3—Since a stoma never opens or closes completely in the dark, is it possible that some additional factor (besides CO_2 concentration), not tested for is involved in controlling stomatal movement?

Answer 3

II

Question 1—What is a possible role of photosynthesis in causing the stomata to open? Is this a direct role or an indirect role?

Answer 1

Question 2—Is it possible that photosynthesis has another effect not tested for?

Answer 2

III

Question 1—Explain why you would expect K^+ to increase initially and then to decrease in guard cells of epidermal strips exposed to CO_2- free air in the dark.

Answer 1

Question 2—Give a sequence of events, corresponding to 1–4, that would explain why stomata, which are open during the day, close at night.

Answer 2

IV

Question 1—If there were one epidermal cell per guard cell, what might happen to this epidermal cell once it gives up its K^+? Why?

Answer 1

Question 2—Recognizing that there are many epidermal cells per guard cell, explain why you would not expect epidermal cells to suffer dehydration due to the passage of K^+ from epidermal cells to the guard cells.

Answer 2

PART III

Evolution and Ecology

23 Case Study: Evolution and Mimicry (Difficulty Level 1)

Natural selection acts to select those organisms that are most "fit" and most likely to survive and reproduce in a particular environment. Selection is not a conscious process on the part of organisms but arises because of random mutations and the expression of characteristics by a population. Those organisms with the characteristics that provide the greatest fitness for the environment are more likely to survive and reproduce. Thus, survival and reproduction are a consequence of possessing favorable characteristics under given conditions.

Selection pressure can act constantly over a long period of time on a population and produce highly unique and unusual results. One proposed consequence of such selection is that organisms evolve that closely resemble other, entirely different organisms or objects if this resemblance provides a distinct survival advantage. In other words, selection can act to produce organisms that mimic other organisms or objects. This proposed phenomenon has been termed *mimicry.*

Although there are fewer proposed cases of mimicry in plants than in animals, several examples of the evolution of mimicry in plants have been investigated and can yield interesting information.

Hypothesis—Mimicry in plants arises through the process of selection and results in a greater survival and reproductive advantage to the species that mimics another species or object.

Although there are a number of examples of plants that *apparently* exhibit mimicry, performing experiments that can test the hypothesis is more difficult. For example, some plants in Africa appear very similar to stones (fig. 23.1). Supposedly, this appearance provides camouflage and protection from predators, but experiments have not been done to test this possibility.

A direct test of the hypothesis has been done with 2 different species of flowers in western Europe. Bellflowers and the red helleborine orchid grow in the same areas. Both are visited by 2 species of solitary bees. Significant differences exist between the species of flowers, however, and any evidence of mimicry is *not* obvious. Bellflowers produce large quantities of nectar, while the orchids do not. Moreover, the bellflowers are violet-blue, while the orchid is red, and the flowers are very different in shape.

In spite of these differences, the solitary bees visit and pollinate both species. It is logical that the bees should visit bellflowers since nectar is available and is gathered by the bees. But why should the bees visit the orchids even though no nectar is available? Is there some characteristic of the orchid that makes it appear similar to the bellflowers?

If the characteristics of the flowers are examined in more detail, what would you predict?

Prediction 1—If mimicry has occurred, some common characteristic should be found between the bellflowers and orchids that make them equally attractive to the bees.

Result 1—When the colors of the flowers were examined in the range observed by the bees, it was found that the flowers of both species appear to be the same color to the bees. Bees see best in the blue-ultraviolet and little, if any, in the red. Therefore, the orchid that appears red to humans is seen as blue-violet (similar to the bellflower) by the bees.

Apparently selection has acted on the orchid to produce flowers that appear to the bees to be similar in color to the bellflower. In other words, the orchid mimics the bellflower's color. The result is that bees visit flowers of both species even though orchids provide no nectar.

An equally important result is that when orchids and bellflowers grow in the same areas, the fruit and seed production by orchids is 6 times greater than when the orchids grow in areas away from bellflowers.

Question—Why is the observation that fruit and seed production is 6 times greater when the orchids and bellflowers grow in the same area an important one?
Answer—This result clearly indicates that orchids growing in the same locality as bellflowers have a reproductive advantage over orchids growing in localities away from bellflowers. Thus, mimicking the bellflowers results in more effective pollination by the bees, greater seed production, and, therefore, greater reproduction.

Figure 23.1 Proposed example of stone mimicry. These plants from southern Africa closely resemble the surrounding stones. Supposedly, this similarity in appearance provides camouflage, which allows escape from potential predators. *(©John Trager/Visuals Unlimited)*

If the shape of the bellflowers and/or orchid flowers were changed, how should this affect pollination (i.e., visits by the bees)?
How could you further test the interpretation that similar flower color attracts bees to both bellflowers and orchids?

Natural selection may not be the only selection that leads to mimicry. Hand-weeding of crops provides intense selection pressure *against* weeds and other plants that appear different from the planted crop. If weeds appear similar to the crop plants, however, they can be overlooked by people weeding the crop and survive. Thus, a similar appearance is a very definite survival advantage to these weeds.

Has selection resulted in mimicry by weeds? Investigations of the barnyard grasses (genus *Echinochloa*) have helped answer this question. These grasses are a serious crop pest in Asian rice fields. Ordinarily, the grasses are distinguishable from young rice plants because the grasses have reddish leaf bases and drooping leaves as shown in figure 23.2

What would you predict about the selection exerted on the barnyard grasses by weeding in the rice field?
Prediction 2—Weeding should select for grasses that are more similar in appearance to young rice plants.

Result 2—Barnyard grasses that closely resemble rice plants have appeared in the rice fields. The characteristic reddish leaf bases and drooping leaves have been lost, and barnyard grasses have evolved that are similar in appearance to rice plants. Apparently, as random mutations in the barnyard grasses give rise to these changed characteristics, these mimic grasses escape detection. Consequently they have spread widely, becoming a serious problem throughout rice-growing regions.

In another example, plant geneticists in India developed rice varieties with purple leaves. Workers weeding the fields could easily distinguish weeds from the purple-leaved rice plants. Within a very short time, however, the weedy wild grasses had hybridized with the cultivated rice, thus exhibiting the purple-leaved character and again escaping detection. The weeding process provided selection favoring the weeds with purple leaves and once again selection had resulted in a mimic plant.

Figure 23.2 Mimicry of young rice plants by barnyard grasses. On the far left is a 3-week-old rice plant and on the far right is a typical barnyard grass with drooping leaves and reddish leaf bases. The middle plant is a barnyard grass that mimics the rice plant (notice that the characteristics of drooping leaves and reddish leaf bases have been lost). *(Courtesy S. C. H. Barrett/University of Toronto)*

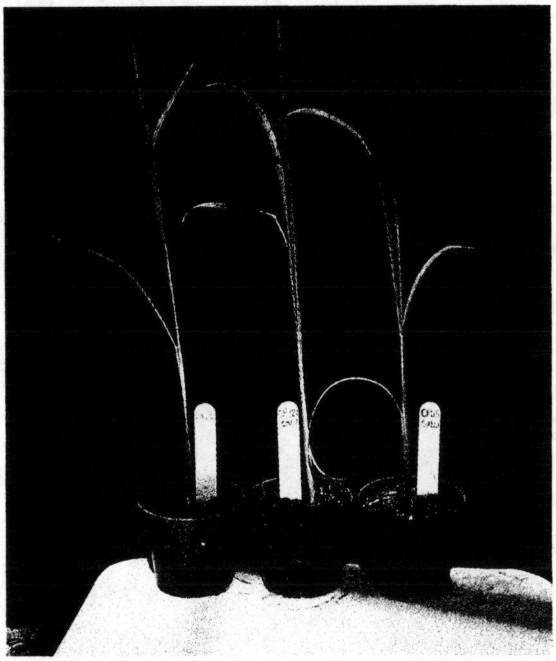

Question—What would happen if random mutations (which result in a similar appearance between the potential mimic plant and the plant being mimicked) did *not* arise?
Answer—If mutations did not arise, there would be no new genetic variability that could be acted on by selection and no change could occur. Preexisting genetic variability would also be acted on by selection, however, and could account for the origin of mimics even without mutations.

II

If the weedy wild grasses could *not* interbreed and hybridize with the cultivated purple-leaved rice, could a mimic have arisen?
If barnyard grasses growing wild in uncultivated areas were observed, what characteristics would they most likely exhibit?
Why is selection necessary to produce a mimic plant?

While there are many cases indicating that selection produces mimicry, it is necessary to demonstrate experimentally that mimicry is a definite survival advantage. These experiments and observations have provided such evidence and represent considerable support for the hypothesis.

References

Barrett, S. C. H. September 1987. Mimicry in plants. *Scientific American,* 76–85.
——. July–September 1983. Crop mimicry in weeds. *Economic Botany,* 255–82.

Case Study 23 EVOLUTION AND MIMICRY

Name ____________________
Section ____________________
Date ____________________

I

Question 1—If the shape of the bellflowers and/or orchid flowers were changed, how should this affect pollination (i.e., visits by the bees)?

Answer 1

Question 2—How could you further test the interpretation that similar flower color attracts bees to both bellflowers and orchids?

Answer 2

II

Question 1—If the weedy wild grasses could *not* interbreed and hybridize with the cultivated purple-leaved rice, could a mimic have arisen?

Answer 1

Question 2—If barnyard grasses growing wild in uncultivated areas were observed, what characteristics would they most likely exhibit?

Answer 2

Question 3—Why is selection necessary to produce a mimic plant?

Answer 3

24 Case Study: Formation of Species (Difficulty Level 1)

Animals or plants within one species can interbreed with other members of the species but do *not* normally interbreed with organisms of another species. The existence of some species can seem puzzling, however, because different species may appear to be very similar but will continue to exist as separate species. For example, some species of ducks live in the same areas, even building their nests close to each other, but are not known to interbreed in their natural environment.

How can clearly separate, yet very similar, species arise and continue to exist? Are new species continuing to form? And if so, how does this happen? These represent major questions asked by biologists and have generated numerous investigations of species formation. In some cases, particularly in plants, polyploidy has been found to generate new species quickly (see Case Study polyploidy). Polyploidy, however, rarely occurs in animals and clearly can only account for a few new species in plants.

Several observations have guided the proposal of a hypothesis to explain species formation.

1. Members of a species freely interbreed. Any genetic changes, such as mutations, that occur in the species are spread to other members of the species.
2. If members of a species are *isolated* so that interbreeding cannot occur, such as by habitation of an isolated island or separation by a mountain range, any genetic change remains within the isolated population. Such changes are mutations that occur in the isolated population and adaptations to local environmental conditions (see Case Study 28, "Genetic Variation and Evolution").
3. Adaptation and mutations are *gradual* so that change in the isolated population occurs very slowly.

These observations have led to the following hypothesis:

Hypothesis—New species can be formed by the isolation of part of an existing species and its adaptation to a new environment. Adaptation will result in sufficient change so that the isolated part of the species cannot interbreed with the original species even if they again inhabit the same environment.

What evidence exists to support this hypothesis? How can it be tested? Based on the hypothesis (and the observations that led to the hypothesis), predictions can be made about what should be found when examining natural populations in areas where isolation is likely to occur. Since adaptation occurs slowly, the following prediction can be made.

Prediction 1—Investigations over a wide geographic area should reveal groups of organisms (of the same species) in various *intermediate* stages of change leading to the possible formation of separate species.

Result 1—Many instances have been found of species at intermediate stages of formation. Several of these are found where islands exist, providing sites for isolation. Some of the best examples are with birds and insects where reproduction is rapid and a small population can be easily isolated. When examining such isolated populations, it is commonly observed that the different groups differ slightly from each other in observable characteristics. These differences may be greater between some groups than others. Moreover, when some groups are brought together, they can interbreed, while others cannot do so.

These observations therefore support the hypothesis by providing examples of genetic change in species that have been split apart and isolated.

Question—Have any of the isolated groups described become new species?

Answer—Those that cannot or do not interbreed when again brought into contact can be considered new species.

If left isolated for a longer period of time, will those groups that could still interbreed if brought together become new species?
Would you expect that new groups are continually becoming isolated?
If a small part of the species were only *partly* isolated (i.e., if occasionally other members of the species could enter or leave the isolated group), could a new species arise?

Figure 24.1 Comparison of beak lengths of 2 species of nuthatches where the living areas overlap (C, D, E, F, G); where only species 1 lives (A, B); and where only species 2 lives (H, I, J, K, L, M).

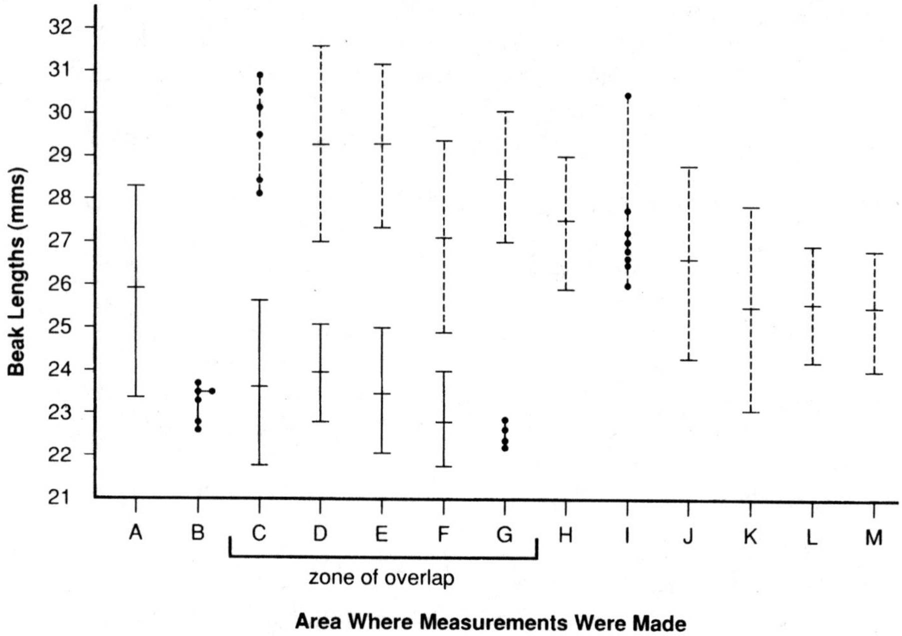

What happens if 2 new species that evolved from the same original species come into contact in the same environment? Since the species are still very similar, they will likely be competing for the same food and resources in the environment. Will such competition lead to further change in the 2 species? It would be expected that adaptation would decrease competition between the species because the more competition between the species for the same food and other resources, the fewer the resources available for either one of the single species. In other words, adaptation should lead to the 2 species exploiting different food sources, living sites, and other resources and, thus, becoming less competitive.
Prediction 2—When 2 new species derived from the same original species live in the same environment, selection and adaptation will act to minimize competition.
Result 2—When examples of 2 new species, probably resulting from the same original species, occupy the same environment, *greater* differences in observed characteristics are found than in areas where the 2 species do not live in the same environment.

For example, the beak size was measured in 2 species of nuthatches whose living areas partially overlap. As shown in figure 24.1, the range of beak length of the 2 species is different when the species occupy the same areas. Beak length overlaps considerably, however, where one species lives without the other.

In another example, 2 species of flowers of the same genus were examined. In areas where the 2 flowers grew together, the flower sizes were very different. In areas where one species grew without the other, however, flower size was less different. This size difference is related to the size of 2 species of bees that pollinate the flowers. Apparently, variations in flower size in areas where they live together represent adaptations to reduce competition for the same pollinators.

These results support the hypothesis by showing that when new species are formed, adaptation and selection continue to act to diverge the 2 new species further.

II In the example of the nuthatches, the difference in beak length probably represents decreased competition for what resource?

When 2 new species are brought together in the same environment, why is it necessary that there be *no* interbreeding if the species are to remain separate?

In the example of the flowers, what advantage is there to decreased competition for pollinators?

The observations and results presented here provide a reasonable and logical explanation for the way in which new species are generated. The investigation into the formation of new species is difficult; biologists must find preexisting cases, such as isolated populations, where predictions can be tested by gathering extensive observations. Moreover, a great deal of data must be collected, which requires observing and collecting organisms over a large area. In spite of these difficulties, however, a great many cases have been investigated that tend to support the hypothesis of species formation.

References

Grant, V. 1981. *Plant Speciation,* 2d ed. New York: Columbia University Press.

Mayr, E. 1976. *Populations, Species, and Evolution.* Cambridge, Mass.: Harvard University Press.

Stebbins, G. L. 1971. *Processes of Organic Evolution,* 2d ed. Englewood Cliffs, N.J.: Prentice-Hall.

Case Study 24 FORMATION OF SPECIES

Name ____________________
Section ____________________
Date ____________________

I

Question 1—If left isolated for a longer period of time, will those groups that could still interbreed if brought together become new species?

Answer 1

Question 2—Would you expect that new groups are continually becoming isolated?

Answer 2

Question 3—If a small part of the species were only *partly* isolated (i.e., if occasionally other members of the species could enter or leave the isolated group), could a new species arise?

Answer 3

II

Question 1—In the example of the nuthatches, the difference in beak length probably represents decreased competition for what resource?

Answer 1

Question 2—When 2 new species are brought together in the same environment, why is it necessary that there be *no* interbreeding if the species are to remain separate?

Answer 2

Question 3—In the example of the flowers, what advantage is there to decreased competition for pollinators?

Answer 3

25 Case Study: Rain Forest Epiphytes (Difficulty Level 1)

In a tropical rain forest, there is intense competition for sunlight, but ample rainfall and favorable temperatures support luxuriant growth. Trees can grow as much as 50 m tall and develop thick canopies that allow only 1% of the sunlight falling on the forest to penetrate to the forest floor. In the struggle for sunlight, heavy vines grow up the trees and can develop such dense foliage that the supporting tree dies as a result of being shaded.

Smaller plants can grow on the branches and trunks of trees and are thus able to grow nearer sunlight. These plants, known as *epiphytes,* are unusual since they can grow high above the forest floor without their roots reaching the ground and without being parasitic on the supporting tree. The roots of epiphytes fasten to the bark of the tree and form thick mats on the trunk or limb surface as shown in figure 25.1.

How do these plants survive? How do they obtain needed water and nutrients without contact with the soil?

Prediction 1—Epiphytes must have specific adaptations that allow them to survive in the conditions under which they live on forest trees.

Result 1—Apparently sufficient nutrients can be obtained from minerals dissolved in rainwater and from material such as falling leaves and twigs that collect in the thick mat of roots and decay to form humus around the roots.

Water is available from rainwater. Even in the rain forest, however, with frequent rainfall and high humidity, obtaining a sufficient supply of water for plants without roots that reach the soil can be a problem. A certain amount of water can be absorbed in the thick mat of roots, and many epiphytes can also trap water in the base of broadened stems, in structures formed by tight rosettes of leaves, and in enlarged stem bases. Thus, a combination of adaptations allow epiphytes to flourish in the rain forest environment.

Question—At what location on a tree would epiphytes be most likely to grow?

Answer—In the junctions where limbs branch from the tree trunk or a crotch formed where the trunk may branch would be particularly favorable positions. Rainwater would naturally tend to run to and collect at these positions. Material, such as leaves and other organic matter, would also be more likely to collect at these positions, providing additional material for humus in the root mat.

I Why would epiphytes more likely be found in a rain forest rather than a typical forest in the northern United States?

Would epiphytes more likely be found on a tree with smooth bark or on a tree with deep cracks and grooves in the bark?

A great diversity of epiphytes can grow on a single tree and will often form a dense mass of growth. Such growth can represent a sizable portion of the living organic material in a forest. Ecologists frequently study the amounts of biomass comprising different plant types or portions of plants. For example, the biomass of leaves and foliage can be measured and compared in different types of plants. Perhaps even more important is the percentage of the nutrients of the ecosystem stored in the various components of the biomass. When these studies are done in rain forests, as much as 40% of the nutrients stored in leaves and other foliage are found to be in epiphytes.

This mass of organic material represents a significant load for a tree to support. On a single tree, the weight of epiphytes can be several thousand pounds. Such a heavy load may break limbs or even lead to the felling of the whole tree. Consequently, epiphyte growth can be a severe disadvantage to a tree, and it has been proposed that at least some rain forest trees have evolved mechanisms to inhibit the growth of epiphytes.

Figure 25.1 An epiphyte growing on the trunk of a rain forest tree. The roots of the epiphyte anchor to cracks and grooves in the bark. Rainwater collects in the pockets formed by the bases of the leaves. *(Asa Wright/Biological Photo Service)*

One mechanism that has been studied is the action of ants living on the *Cecropia* tree. It has been hypothesized that these ants, of the genus *Azteca,* remove epiphyte seeds and young epiphyte plants, and in this way, prevent the dense mass of growth from becoming established. Since the *Cecropia* has fragile limbs that could break under the load, the removal of young epiphytes represents a significant survival advantage to the tree.

Do the ants actively remove foreign plant material from the tree? How could this question be investigated?

Prediction 2—If ants actively remove material, then foreign plant material that is placed on or fastened to the *Cecropia* tree should be removed by the ants.

Result 2—Small mats containing epiphyte plants of varying sizes were tied to *Cecropia* limbs. The ants immediately pulled loose small bits of material from the mats and dropped them off the tree. Larger epiphytes in the mats, however, were not removed. Apparently, the ants can prevent epiphyte growth on the tree, but removal must be done before the epiphytes become established as large plants.

Question—Could the ants harm the growth of larger epiphytes even though they cannot remove the plants?
Answer—The ants do remove parts of the mat in which the epiphytes are growing. The humus and other mat material provide nutrients and absorb and hold rainwater near the roots. Removal of this material could deprive the plant of nutrients and decrease the water available to the plants.
Question—Why should the ants have evolved the behavior of protecting the *Cecropia* from epiphytes? Do the ants have something to gain from this activity?
Answer—Observations of ants living on the *Cecropia* have indicated that the ants benefit from the tree in 2 ways. First, the tree produces nodules containing starch at the base of the leaves and the ants feed on these nodules. Second, the hollow stems of the tree are utilized by the ants for shelter and nesting sites. Thus, protection of the tree by the ants serves to preserve this beneficial environment for the ants.

II Some trees periodically shed sections of their bark. How would this protect the tree against the excessive growth of epiphytes?
Why would the *Cecropia,* in particular, benefit from the protection by ants while many other trees might not?
Could you propose some factors, in addition to epiphyte size, that might protect epiphytes from removal by the ants?

Rain forest epiphytes are an interesting example of an organism that has adapted to a diverse and challenging environment. These plants have developed a unique strategy that allows them to compete successfully for sunlight despite a number of disadvantages. The complex interactions of the forest ecosystem are well illustrated by the epiphyte adaptations and the response of other forest organisms to epiphyte growth.

Reference

Perry, D. R. November 1984. The canopy of the tropical rain forest. *Scientific American,* 138–47.

Case Study 25 RAIN FOREST EPIPHYTES

Name ____________________
Section ____________________
Date ____________________

I

Question 1—Why would epiphytes more likely be found in a rain forest rather than a typical forest in the northern United States?

Answer 1

Question 2—Would epiphytes more likely be found on a tree with smooth bark or on a tree with deep cracks and grooves in the bark?

Answer 2

II

Question 1—Some trees periodically shed sections of their bark. How would this protect the tree against the excessive growth of epiphytes?

Answer 1

Question 2—Why would the *Cecropia,* in particular, benefit from the protection by ants while many other trees might not?

Answer 2

Question 3—Could you propose some factors, in addition to epiphyte size, that might protect epiphytes from removal by the ants?

Answer 3

26 Case Study: Lyme Disease (Difficulty Level 1)

When a new disease is discovered, several questions are immediately asked. What causes the disease? Is it a virus or other microorganism? How is the disease spread? At first, it may not be apparent that the disease is a new one, so the first question would be, Is this a new disease?

These questions were asked about Lyme disease when it was first observed in 1975. The first cases were in eastern Connecticut around areas known as Lyme and East Lyme. Several adults and children reported symptoms characteristic of juvenile rheumatoid arthritis. Ultimately, the affected individuals developed joint problems characteristic of rheumatoid arthritis and in some patients, nerve and muscle difficulties were found. There were immediate indications, however, that this disease outbreak was unusual. First, juvenile rheumatoid arthritis is rare, but in the infected areas, the occurrence was as much as 100 times what would normally be expected. Second, the distribution of affected individuals within the infected area was unusual; most victims lived near heavily wooded areas and only a few in towns away from wooded sections.

When faced with unusual conditions such as these, how do the investigators of a disease proceed? At this time, it was not obvious that a new disease was being observed but only that a disease outbreak with unusual conditions was present. Investigations proceeded with the following hypothesis:

Hypothesis—The outbreak was a new disease with a distinctly different cause.

The first step in an investigation is to make extensive observations on the history of the infection in the area (i.e., when it occurs, who is infected) and the specific nature of the disease symptoms. These procedures were used in the Lyme and East Lyme areas. Several important observations were made:

1. The disease was not very contagious. Many times, an individual in a family would have the disease, yet it would not spread to other family members.
2. In most cases, the disease appeared in the summer only.
3. Many people who had the disease remembered initially having a skin rash before other symptoms appeared. The rash started as a small bump on the chest, abdomen, or back and gradually expanded to cover a large area.

Since none of these observations were characteristic of juvenile rheumatoid arthritis, the investigators began to suspect that they were dealing with a previously unknown disease. Observation 3 led investigators to suspect that the disease was spread by some crawling insect, spider, or tick. This last point is an interesting deduction. Why do you think it was made by the investigators?

Question—Why did the investigators think that the disease was spread by a crawling insect, rather than a flying one?

Answer—The initial small bump suggested a bite. The appearance on the chest, abdomen, and back indicated a bite by an organism that could crawl under the clothes. A flying insect would most likely bite on an exposed part of the body.

Obtaining the evidence to substantiate these suspicions was a more difficult matter. Several months later in subsequent interviews with infected people, one person remembered being bitten by a tick, which he had saved. The tick was identified as *Ixodes dammini,* a very small organism about the size of a pinhead.

How could further investigations be made to decide whether these ticks were the actual carrier of the disease?

The investigators set out traps on both sides of a river running near the infected area. Captured animals could be checked for the presence of ticks. Lyme and East Lyme, where a high incidence of

the disease was found, were on one side of the river. Almost no cases had been reported from people living on the other side.
Prediction 1—If *I. dammini* is found to be present in greater numbers in areas where the disease was found, then this would be a strong indication that the tick is linked to the presence of Lyme disease.
Result 1—*I. dammini* was found to be 12 times more prevalent on the side of the river where Lyme and East Lyme were located. This result provided strong supporting evidence that the disease was carried and transmitted by the tick.
Question—Why did the investigators set out traps on *both* sides of the river?
Answer—Simply finding a high incidence of *I. dammini* on the same side of the river as the disease would not establish a relationship between the tick population and the disease incidence. It was important to find that when the tick population was low, the incidence of the disease was also low.
Question—Why were the 2 sides of the river an obvious location to examine for the presence of ticks?
Answer—The river provides a natural barrier that separates tick populations and prevents the movement of animals carrying the ticks. The presence of the river explains why the disease had not spread (at least not as seriously) to people on the other side and represented an obvious location to test the relation between tick presence and the disease.

I Why would a higher population of ticks lead to a higher incidence of the disease?
Why does the early observation that more people getting the disease lived near wooded areas seem logical knowing that the disease is probably spread by tick bites?

Although these observations were a strong indication the tick carried the disease, far more positive proof would be provided if the bacterium or virus that caused the disease could be found in the ticks. Subsequent investigations found large numbers of an unusual, irregular-shaped bacterium in the digestive tract of the tick.

Is this the microorganism that causes Lyme disease? This question can be examined by extracting antibodies from the blood of people who have been infected by Lyme disease and observing the reaction of the antibodies with the bacteria. If antibodies are present for the specific bacterium causing Lyme disease, they will combine with the bacteria, producing a reaction that can be easily observed. People who have been infected with a disease normally generate antibodies to the infecting organism. These antibodies confer some resistance to future infections. For example, people who have had chicken pox are resistant to another infection of chicken pox due to antibodies now present in their bodies.

What do you predict would be found in this test?
Prediction 2—If the bacterium found in the digestive tract of the tick causes the disease, antibodies from previously infected people should react with this bacterium.
Result 2—Subsequent observations showed that the predicted reaction did occur. Similar tests for antibodies to over 200 known diseases carried by insects, ticks, or spiders did *not* indicate antibodies to any of these diseases.
Question—Why was the observation that antibodies from Lyme disease victims did not react with known disease organisms important?
Answer—This was strong evidence that Lyme disease was a new and previously unknown disease.

II What would be found if antibodies from people who had *not* had Lyme disease were exposed to the bacteria from the ticks?
If a previously uninfected person were injected with the bacteria from the tick, what would you expect to happen?

These investigations gave strong support for the hypothesis that a new disease had been discovered and were helpful in determining the way this disease was spread and how it was caused. Lyme disease has since been found in a larger area of the northeast United States, the northern Midwest, and parts of the Northwest. Although it is a serious concern in these infected areas, the disease can be effectively treated with modern antibiotics if detected early.

Reference

Habicht, G. S., Beck, G., and Benach, J. L. July 1987. Lyme disease. *Scientific American,* 78–83.

Case Study 26 LYME DISEASE

Name ____________________
Section ____________________
Date ____________________

I

Question 1—Why would a higher population of ticks lead to a higher incidence of the disease?

Answer 1

Question 2—Why does the early observation that more people getting the disease lived near wooded areas seem logical knowing that the disease is probably spread by tick bites?

Answer 2

II

Question 1—What would be found if antibodies from people who had *not* had Lyme disease were exposed to the bacteria from the ticks?

Answer 1

Question 2—If a previously uninfected person were injected with the bacteria from the tick, what would you expect to happen?

Answer 2

27 Case Study: Extinction of the Dinosaurs (Difficulty Level 2)

What killed the dinosaurs that roamed the earth millions of years ago? People have long been fascinated with these strange animals and have wondered why they became extinct. It is even more puzzling that extinction occurred over a relatively short time about 65 million years ago, not only for the dinosaurs but also for half the species of plants and animals living at the time. Whatever the cause of extinction, it must have been an incredible catastrophe to have destroyed so much of the earth's life.

Two hypotheses have been proposed to explain the catastrophic events that could have caused the mass extinction:

Hypothesis 1—A giant comet or asteroid crashing into the earth.

Hypothesis 2—Extensive volcanic eruptions.

Both hypotheses predict that enormous amounts of dust and ash would be released into the earth's atmosphere, which would block sunlight and cause major climatic changes. Photosynthesis would have been disrupted, resulting in the widespread death of plants and animals in the food chains.

Is there evidence that either of the catastrophic events proposed by the hypotheses actually occurred?

Prediction 1—If an asteroid impact or major volcanic activity occurred, then there should be some obvious evidence in the earth's crust.

Result 1—In a thin layer of sediment separating the Cretaceous and Tertiary geologic periods (K-T boundary) was found a very high concentration of the element *iridium.* This element is relatively rare in the earth's crust but is much more common in extraterrestrial material, such as asteroids. An impact would have vaporized the asteroid and spread iridium-rich dust around the world. As this dust settled out of the atmosphere, it would form a layer of sediment. It has also been found, however, that iridium-rich material has been produced by volcanic eruptions. This material could have been present in worldwide volcanic dust clouds.

The iridium-rich sediment layer provides evidence that some catastrophic event occurred 65 million years ago that could have led to worldwide extinctions.

Question—Which of the 2 hypotheses are most strongly supported by Result 1?

Answer—Since the observations are consistent with both hypotheses (i.e., iridium-rich dust can be produced by both asteroids and volcanoes), neither hypothesis seems more likely than the other.

What evidence would provide support for one hypothesis over the other? More direct evidence of an asteroid impact on the earth's surface or volcanic activity would be necessary.

Prediction 2—Remains of an asteroid impact on the earth's surface would support the impact hypothesis (2); while remains of extensive volcanic activity would favor the second hypothesis if the events occurred 65 million years ago.

Result 2—The remains of massive lava flows have been discovered covering large parts of India. Moreover, fossils from about 65 million years ago are found in layers between the lava flows, but layers above the flows do not have fossils.

The remains of many impact craters caused by extraterrestrial bodies have been found on earth, but these either do not appear large enough to have caused worldwide extinction or do not occur at the appropriate date.

Based on the evidence in Result 2, which hypothesis would now have the strongest support? Scientists favoring the impact hypothesis have argued that an impact could have occurred in the oceans, and the remains would not be obvious. Would this argument affect the strength of either hypothesis?
Why is dating the catastrophic event so important in evaluating Result 2?

Other observations have indicated the occurrence of some catastrophic event. Geologists have discovered grains of quartz at many locations on the earth's surface. These grains bear characteristic striations indicating that the grains were produced by a great shock (fig. 27.1). Such grains are commonly found near known impact craters and at underground nuclear test sites.

What would you predict about the location of these "shocked" quartz grains?

Figure 27.1 Quartz grain shows laminar deformations generated by hypervelocity shock. Such quartz is found in laboratories, at nuclear test sites, and near impact craters. *(U.S. Geological Survey)*

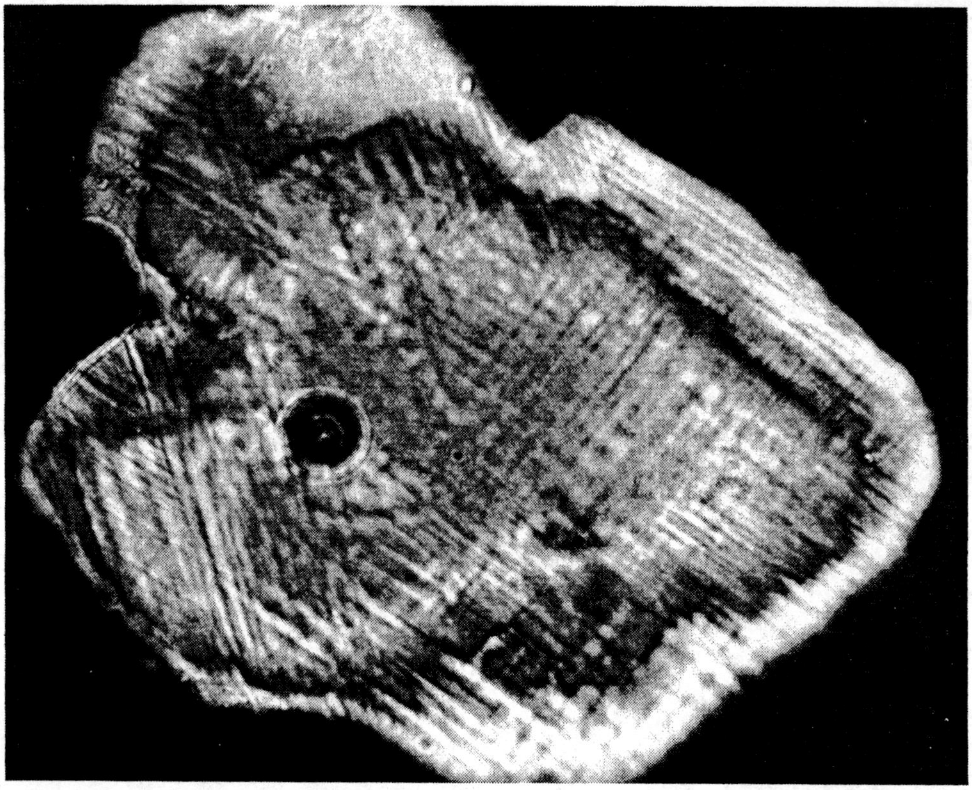

Prediction 3—The "shocked" quartz grains should be found in the K-T boundary if a major impact occurred at the time of extinction.
Result 3—The characteristic "shocked" quartz grains *are* found in abundance in the K-T layer, as well as at impact craters. Supporters of the volcanic activity hypothesis argue that the geologic shocks produced by volcanoes *could* produce "shocked" quartz grains.

II If volcanic activity were responsible for the mass extinctions, what observations would you predict based on Result 3?
Which hypothesis would you now argue as the most likely? What would be your reasons?

Scientists continue to investigate actively the causes of the mass extinction. New evidence is discovered at a rapid rate and before long, perhaps the cause of the great natural catastrophe 65 million years ago will be better understood. The competing hypotheses to explain this event and the evidence gathered to test and evaluate these hypotheses clearly illustrate the process of scientific investigation and the questions scientists ask.

References

Alverez, W., and Asaro, F. October 1990. An extraterrestrial impact. *Scientific American.*
Courtillot, V. E. October 1990. A volcanic eruption. *Scientific American.*

Case Study 27 EXTINCTION OF THE DINOSAURS

Name ____________________
Section ____________________
Date ____________________

I

Question 1—Based on the evidence in Result 2, which hypothesis would now have the strongest support?

Answer 1

Question 2—Scientists favoring the impact hypothesis have argued that an impact could have occurred in the oceans, and the remains would not be obvious. Would this argument affect the strength of either hypothesis?

Answer 2

Question 3—Why is dating the catastrophic event so important in evaluating Result 2?

Answer 3

II

Question 1—If volcanic activity were responsible for the mass extinctions, what observations would you predict based on Result 3?

Answer 1

Question 2—Which hypothesis would you now argue as the most likely? What would be your reasons?

Answer 2

28 Case Study: Genetic Variation and Evolution (Difficulty Level 2)

Both natural selection and genetic variation in populations are essential components of the modern theory of evolution. Genetic variation presents a diversity of characteristics on which natural selection can act to favor the survival of those characteristics that are advantageous to a population. In this way, gradual change over many generations results in adaptations that uniquely suit organisms to their environment.

While this generalized scheme provides an overview of the mechanism of evolution, several critical questions can be immediately raised. How much genetic variation exists in populations? Is there a relation between genetic variation and the rate of evolutionary change?

Such questions serve to generate testable hypotheses on the relationship between genetic variation and the process of evolution. One such hypothesis can be stated as follows:

Hypothesis—Natural populations contain a large amount of genetic variation. Selection acts on this large pool of genetic variation to produce adaptive change.

In an initial test of this hypothesis, 2 populations of fruit flies were subjected to intense selection pressure. One population (A) was a single pure strain, while the other population (B) was produced by crossbreeding 2 single strains.

Due to the crossbreeding, population B contained about twice as much genetic variation as population A. Equal numbers of flies from these 2 populations were then placed in separate large culture bottles, where the flies would be subjected to intense competition for food and space. After 500 days (about 25 generations), the number of flies in each bottle was counted. What result would you predict?

Prediction 1—The greater genetic variation due to crossbreeding to initially produce population B should permit more rapid adaptation to the environment. Therefore, more flies should be found in population B than in A after 25 generations of selection.

Result 1—The results of this experiment are shown in figure 28.1. Although both populations increased in number, population B (with more genetic variability) increased more rapidly. This result suggests that population B adapted to the environmental conditions more rapidly. These results clearly indicate that selection pressure leads to adaptive change and supports the hypothesis. Moreover, the greater the genetic variability, the more rapid the adaptation.

Figure 28.1 Increase in number of flies maintained in separate culture bottles over 500 days (about 25 generations). Population A is a single pure strain. Population B was produced by crossbreeding 2 single strains and contains about twice as much genetic variability as population A.

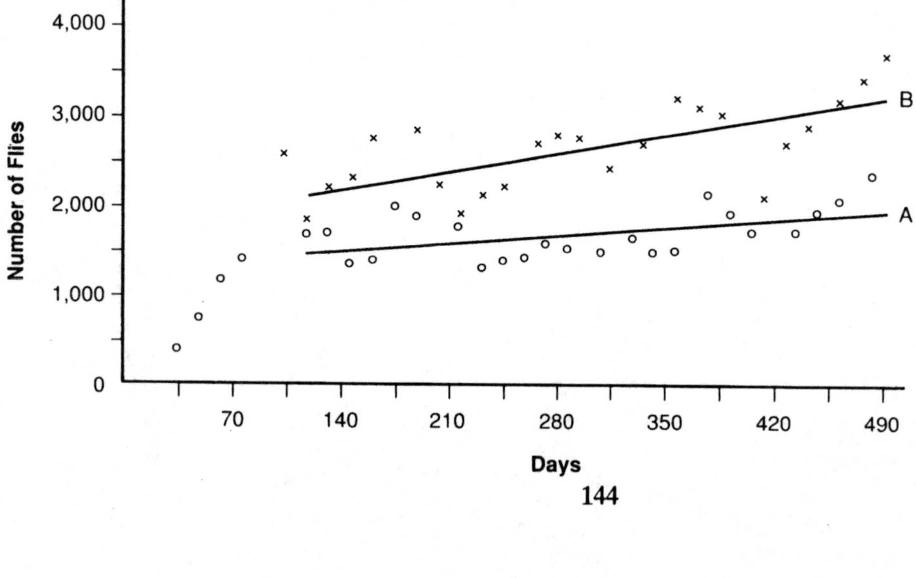

Question—Why does this experiment indicate adaptive change rather than just the growth of the population?
Answer—The fly populations gradually increased over 25 generations. Population growth (without adaptive change) should quickly (within 2–3 generations at most) reach the maximum population size capable of being supported by the environmental conditions and then remain stable. In this experiment, however, the populations continued to increase slowly over many generations. This slow growth indicated selection was producing adaptation. Each successive generation was genetically more "fit" for the environment. As a result, greater numbers survived in each successive generation.

I

If flies from population A were bred with flies from population B, would the offspring be *more* or *less* able to survive in the culture bottles?
What were the selection pressures in this experiment?

While this experiment does show adaptation, it does not indicate selection of specific traits or show how selection occurred. Can highly focused selection lead to specific changes?

In another experiment, selection was artificially controlled and changes in specific traits observed. Corn kernels were artificially selected for either high or low oil content. In other words, experimenters selected those corn kernels in each generation that were highest in oil content and, also, those that were lowest in oil content. High oil and low oil kernels were planted in separate plots. The resulting plants in the separate plots were not allowed to interbreed. These plants produced the next generation. This procedure was continued for 50 generations. Would this highly directed selection for a specific trait lead to genetic change?
Prediction 2—If the genetic variation for oil content is sufficiently large, changes in oil content should continue in response to selection for many generations.
Result 2—The results of this experiment indicated oil content increased from 5% to 14% and was still increasing even after 50 generations of intense selection for the group with high oil content. In the group selected for low oil content, no significant change was observed in the later generations while oil content had decreased from 5% to less than 2% in the early stages. These results provide further support for the hypothesis by showing selection can lead to highly significant change in specific traits.

II

What if interbreeding occurred between high oil and low oil corn? What would you predict about the oil content of the offspring?
Would these offspring have more or less genetic variability for oil content compared to the high and low oil content parents?
If reverse selection were applied to the high oil corn (i.e., if kernels were now selected for *low* oil content), what would you predict would be observed?

The evidence from these experiments suggests that natural populations *do* contain a large amount of genetic variation but provides no measure of how much variation exists. New techniques in molecular biology, however, have allowed much more extensive and precise measurements. A particularly useful technique is the electrophoresis of proteins.

In electrophoresis, an electric field is applied across a gel plate composed of starch or synthetic polymer, with material, such as a protein, placed on the gel. If the molecules of the material have an electrical charge, the molecules will move across the plate in response to the electrical field. The distance of movement in a specified time period is related to the charge on the molecule. The greater the charge, the further the movement.

Since the amino acids composing the proteins in an organism are coded by the sequence of nucleotides in DNA, any change in the DNA should be reflected by a change in the amino acid sequence of the protein. Such changes can be detected by electrophoresis of proteins because different amino acids carry a different electric charge. Proteins with different electric charges will travel different distances when subjected to electrophoresis. These different distances of travel will separate proteins on the electrophoresis plate and indicate very slight genetic changes in proteins (i.e., changes in the DNA).

This technique is especially useful in detecting slight differences existing in proteins that are synthesized from the same gene on homologous chromosomes. Each diploid cell contains homologous pairs of chromosomes. The pair of homologues contain the same genes, but mutation may have led to differences between the DNA of the same gene on the 2 homologues. For example, suppose the gene for a specific enzyme is carried on both homologues. It is the same gene but the amino acid sequence may differ slightly in the enzyme produced by each homologue, even though both forms are still functional and catalyze the same reaction. If these differences are reflected by differences in the electric charge, the 2 enzyme forms can be separated by electrophoresis.

What would you predict would be observed if a specific enzyme were examined in several individual organisms of the same species?

Prediction 3—Genetic variability existing in the DNA coding for the enzyme will be detected in the electrophoretic movement of the enzyme in tissue from separate organisms.

Result 3—Figure 28.2 shows the electrophoretic separation of a specific enzyme from 10 fruit flies of the same species. Three of the individuals have one form of the enzyme, while 7 possess a second form. Obviously, the different forms of this enzyme represent a high degree of genetic variability.

Electrophoretic analysis of a great many proteins from numerous species of animals has indicated that for vertebrates, the average percentage heterozygosity (the percentage of genes that are dissimilar) is 6.6% for each individual. In other words, the average individual is heterozygous for 6.6% of their genes. Investigators consider such figures a large amount of genetic variability. Indeed, these figures were much higher than many geneticists had suspected existed before the application of new experimental techniques in molecular biology.

Figure 28.2 Electrophoresis of proteins from 10 individual fruit flies. Tissue samples were placed at the bottom of the plate and an electric field applied between the top and bottom of the plate for a period of time. The final positions of 2 forms of a specific enzyme are indicated by applying a stain specific for the enzyme being investigated. Three of the flies have one form of the enzyme and 7 have the other form.

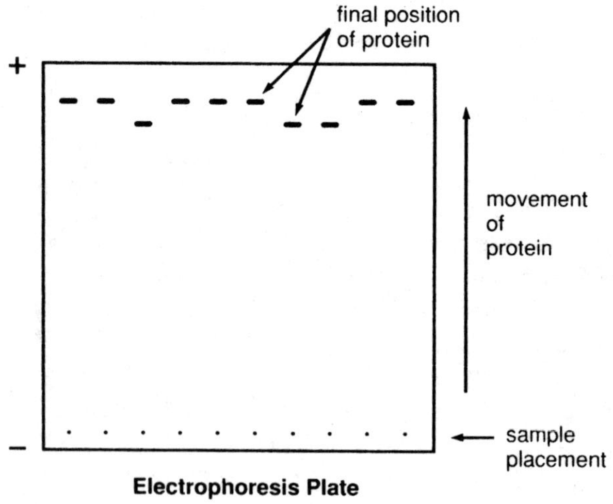

III Could 3 or more forms of a protein from the same gene be found in a population of organisms?
In Experiment 1, which population should show the greatest percentage heterozygosity?
Why should intense selection result in decreased heterozygosity?

Diverse studies involving both natural and artificial selection and investigations of the molecular structure of proteins have been used to study genetic variability and the process of evolution.

Scientists investigating evolution and genetic variability consider these experiments strong support for the hypothesis that populations contain large amounts of genetic variation and that selection acts on this variation to bring about adaptive change.

References

Ayala, F. J., ed. 1976. *Molecular Evolution.* Sunderland, Mass.: Sinauer Associates.

—. September 1978. The mechanisms of evolution. *Scientific American,* 56–69.

Case Study 28 GENETIC VARIATION AND EVOLUTION

Name ____________________
Section ____________________
Date ____________________

I

Question 1—If flies from population A were bred with flies from population B, would the offspring be *more* or *less* able to survive in the culture bottles?

Answer 1

Question 2—What were the selection pressures in this experiment?

Answer 2

II

Question 1—What if interbreeding occurred between high oil and low oil corn? What would you predict about the oil content of the offspring?

Answer 1

Question 2—Would these offspring have more or less genetic variability for oil content compared to the high and low oil content parents?

Answer 2

Question 3—If reverse selection were applied to the high oil corn (i.e., if kernels were now selected for *low* oil content), what would you predict would be observed?

Answer 3

III

Question 1—Could 3 or more forms of a protein from the same gene be found in a population of organisms?

Answer 1

Question 2—In Experiment 1, which population should show the greatest percentage of heterozygosity?

Answer 2

Question 3—Why should intense selection result in decreased heterozygosity?

Answer 3

29 Case Study: Testosterone and Aggression (Difficulty Level 2)

Aggressive behavior is often observed in male birds when territories are being established and defended during the mating and nesting season. It has also been observed that testosterone, a male sex hormone secreted by the testes, can increase aggression. Is there a relationship between the observed aggression and testosterone levels in the blood of the birds? If there is a relationship, are there specific environmental signals or cues that can affect testosterone level?

These observations and questions have led to the following "challenge" hypothesis:

Hypothesis: Testosterone and aggression are related only during times when males are in active conflict.

This hypothesis can be tested by monitoring the level of testosterone in birds' blood during periods when birds exhibit frequent aggression and at other periods when aggression is less frequent. A small amount of blood can be periodically removed from a wing vein for testing testosterone levels.

What do you predict would be observed in this test?

Prediction 1—If testosterone level and aggression are related, then blood testosterone level should be higher in birds when aggression is being exhibited.

Result 1—The results of an experiment with the song sparrow are shown in figure 29.1. Two sharp and distinct peaks of testosterone level are observed. The first occurs when a territory is being established. The second peak corresponds with the egg-laying period, when the male is guarding the female. Aggressive male behavior is exhibited during both these periods and is indicated by the dark bars at the bottom of the graph. These results are consistent with the prediction and support the hypothesis.

Question—Do the results indicate that testosterone *causes* aggressive behavior or that aggressive behavior increases testosterone level?

Answer—It is difficult to establish any cause and effect relationship from these data because increases in aggression and testosterone level correspond so closely. If aggression had clearly increased *before* the increase in hormone level, then the indication would be that aggression causes the release of testosterone.

I What if the data used to construct the graph in figure 29.1 were collected infrequently? Would the shape of the graph and the information derived from it change?

The relationship between aggression and testosterone was further investigated in the following experiment. A male song sparrow was removed from his established territory, and within 12–72 hours another male (replacement male) moved into this vacant territory. There was a resulting increase in conflict between this replacement male and the neighboring male song sparrows over establishment of the territorial boundaries.

Figure 29.1 Blood testosterone level in the male song sparrow.

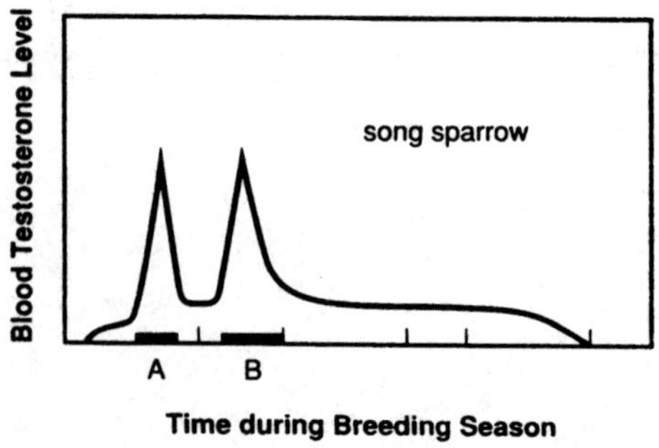

What would you predict about the blood testosterone levels in the replacement male and the other males in conflict with him?

Prediction 2—Testosterone levels in males engaged in conflict should increase. The data in figure 29.1 indicate that testosterone level falls after a territory is established. It could also be predicted then, that males from more distant, undisturbed, and established territories (where there is little if any contact between neighboring males) should show relatively low testosterone levels.

Result 2—Measurements of the blood testosterone levels in the conflicting males indicated, as predicted, significantly increased testosterone levels, while males in undisturbed, established territories showed *no* increase.

Question—Why was it important also to measure testosterone levels in males from distant established territories who were not in conflict?

Answer—These males from established territories serve as controls to examine what happens to blood testosterone in a similar environmental situation at the same time period but without conflict. Without data from these controls, the effect of conflict could not be separated out, and it could then be argued that testosterone is increased in *all* birds occupying territories during this time period.

II What do you expect to happen to blood testosterone in the replacement male *after* his territory is established?

This experiment still leaves open the question of the timing of testosterone increase and aggressive behavior. Does testosterone level increase before or after aggression? To investigate this question, a male intruding on another song sparrow's established territory was simulated by placing a caged male within the territory and playing recorded song sparrow bird songs from speakers near the cage. The established male attempted to attack the bird in the cage and drive him from the territory.

What do you predict would happen to the blood testosterone level in the established male?

Prediction 3—Based on Result 2, testosterone level will increase, but the timing of this increase (before or after the actual attack on the other bird) cannot be predicted with confidence.

Result 3—When testosterone levels were sampled frequently and monitored closely, it was found that testosterone did not increase until after about 10 minutes of aggressive behavior. This result clearly indicates that aggression precedes blood hormone increase. If the caged bird was removed in less than 10 minutes (and the established male engaged in aggressive behavior for less than 10 minutes), no hormone increase could be detected in the established male.

III If it takes 10 minutes for an increased testosterone level to be detected, what do these results indicate about the length of conflicts between males *after* territories are established?
During the establishment of territories?
How would you design an experiment to test whether the established male in this experiment was responding to the appearance of the invading male, the recorded bird song, or both?

These experiments illustrate how relatively complex behavior can be investigated and how the timing of this behavior is related to physiological responses. A number of other questions could be asked. For example, What does testosterone do? Does it affect nerve activity or brain function? Investigation of such questions would require considerably different experimental designs.

Reference

Wingfield, J. C., et al. November–December 1987. Testosterone and aggression in birds. *American Scientist,* 602–8.

Case Study 29 TESTOSTERONE AND AGGRESSION

Name ____________________
Section ____________________
Date ____________________

I

Question 1—What if the data used to construct the graph in figure 29.1 were collected infrequently? Would the shape of the graph and the information derived from it change?

Answer 1

II

Question 1—What do you expect to happen to blood testosterone in the replacement male *after* his territory is established?

Answer 1

III

Question 1—If it takes 10 minutes for an increased testosterone level to be detected, what do these results indicate about the length of conflicts between males *after* territories are established?

Answer 1

Question 2—During the establishment of territories?

Answer 2

Question 3—How would you design an experiment to test whether the established male in this experiment was responding to the appearance of the invading male, the recorded bird song, or both?

Answer 3

30 Case Study: Air Pollution and Forest Ecology (Difficulty Level 2)

Forests in several parts of the Northern Hemisphere have exhibited widespread tree damage leading to gradual forest decline and ultimately the widespread death of many trees. In parts of Germany, forest decline is especially apparent, and these areas have been studied in an attempt to discover the causes of the damage. The most obvious symptom of damage has been needle yellowing and needle loss in conifers.

Air pollution has often been suggested as a cause of the observed forest decline. Pollutants such as sulfur dioxide (SO_2), nitrous oxide (NO_x), ozone, and ammonium (NH_3) can result in acid rain and direct damage to forest trees. In the areas of Germany exhibiting the most extensive damage, the concentrations of SO_2 in the air are well above levels that cause damage, while NO_x and ozone can reach potentially harmful levels. Based on these observations, the following hypothesis has been proposed:

Hypothesis—Airborne pollutants (SO_2, NO_x, and ozone) cause needle yellowing, needle loss, and long-term forest decline.

In one test of this hypothesis, the rate of photosynthesis was compared in needles of the Norway spruce at a declining forest site in Germany (using trees with as yet no obvious signs of damage) and at an air pollution—free site in New Zealand. What results would you predict from this comparison?

Prediction 1—If airborne pollutants have a damaging effect on the spruce needles, then the rate of photosynthesis in the needles should be lower at the declining site in Germany.

Result 1—The results of the comparison in spruce needles of different ages are shown in figure 30.1. Contrary to the prediction, the airborne pollutants showed *no* damaging effects on photosynthesis. Although photosynthetic rate decreased slightly with age, the same effect was observed at both sites.

Question—Why was the comparison of trees at 2 different sites a good experimental design?

Answer—The experimental design allows comparison of trees exposed to a polluted environment (Germany) with trees *never* exposed to significant levels of pollution (New Zealand). If trees from a forest in Germany that appeared healthy were used as controls in this comparison rather than trees in New Zealand, then it might be argued that these so-called healthy trees had been exposed to pollution to some extent and had already exhibited a decline in photosynthesis.

Question—In what ways could this comparison of the 2 sites be misleading?

Answer—In such completely different sites, the trees would be exposed to many uncontrolled variables such as different organisms (plant and animal) in their environments and different soil and weather conditions.

Figure 30.1 A comparison of the rate of photosynthesis in needles of different ages in trees at 2 sites. The rate of photosynthesis is measured as the rate of CO_2 uptake.

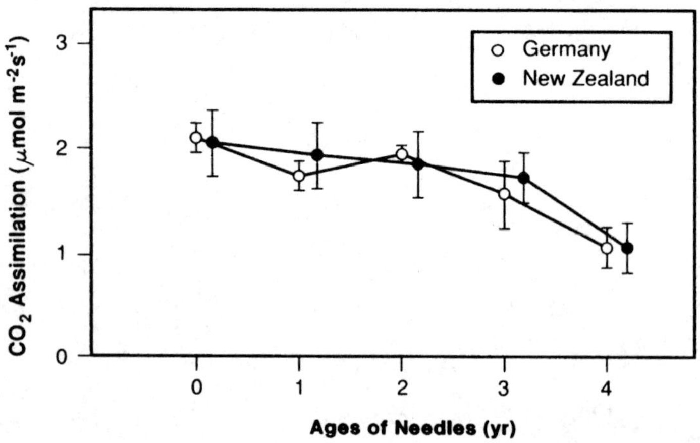

I What other comparisons might have been made between trees at the 2 sites?
Why was the photosynthetic rate examined in needles of varying age?

Other observations have indicated that tree *growth* and air pollution may be strongly related to the forest decline symptoms. In one experiment using spruce trees that were beginning to show needle yellowing, the buds were removed from some twigs in the trees so that no extension growth would occur. The condition of the needles on these twigs was compared to needles on other twigs of the same tree with intact buds. Thus, no growth would occur on twigs with buds removed, while twigs with intact buds would grow in length. What results would you predict?

Prediction 2—If tree growth and air pollution interact in the forest decline symptoms, then twigs that do *not* grow should be less affected by air pollution than twigs that continue to grow.

Result 2—By the end of the subsequent growing season, the twigs with buds removed had regained the green color characteristic of healthy needles. The twigs with buds that exhibited continued growth, however, remained yellow. These results, as proposed by the hypothesis, suggest that air pollution *does* influence forest decline, but this effect is also somehow related to growth.

Question—What is the appropriate control for this experiment?

Answer—The effect of removing buds and of leaving buds intact on trees that are *not* exposed to air pollution should be observed. This control would indicate the effect of bud removal on needles of healthy trees. When this was done, needle color and general health was the same on the 2 twigs (the one without a bud, and the other with the bud intact).

II What are the advantages in this experiment of using twigs on the same tree?
Why is it necessary to check the effect of removing buds in *both* healthy trees and in trees showing evidence of damage?

Why should growth be related to air pollution damage? One possibility is that growth requires additional use of nutrients and minerals. Perhaps under conditions of air pollution, trees are unable to obtain sufficient nutrients to support additional growth but are able to maintain the health of the tree if growth does not occur.

To investigate further the effects of pollution on nutrient uptake, changes in forest soils were studied. Even though SO_2, NO_x, and NH_3 are nutrients that are often used as fertilizers, the input of these substances into the soil from air pollution can result in harmful effects such as soil acidification. When these soils become sufficiently acidic, the availability of Mg, Ca, and K in the soil can be inhibited.

In one experiment, spruce seedlings were grown in acid soils from forest decline areas. These seedlings were then watered with solutions containing ammonium sulfate (which would be present in rain in areas where air pollution was present). What do you predict would be observed in this experiment?

Prediction 3—If substances from air pollution entering the soil affect nutrient uptake, then these seedlings should exhibit nutrient deficiencies.

Result 3—The seedlings in this experiment showed symptoms of Mg deficiency. Apparently, the ammonium sulfate can inhibit Mg uptake in acidic soils.

In more mature trees, the Mg demands of new growth are satisfied by translocating Mg from older needles to the new growth, thus leading to yellowing of the older needles. A complicating factor is that the nitrogen needs of the tree are supplied by the nitrogen compounds in air pollution. Thus, the nitrogen input *stimulates* growth, but a shortage of other minerals (i.e., Mg) results in deficiency symptoms *due* to enhanced growth.

III Why would air pollution damage not be as obvious if growth were slower?
How would you explain Result 2 now after knowing Result 3?
What other experiments could be performed to provide more information on nutrient uptake and availability.

These results support the hypothesis by indicating that airborne pollutants lead to needle damage and forest decline. The damaging effects, however, are frequently exerted indirectly. Clearly, the effects of airborne pollutants are complex and can interact in a variety of ways. Such effects include soil acidification, changes in soil nutrient availability, and the inhibition of nutrient uptake by roots.

A more detailed understanding of these effects, based on continuing research, will suggest control and remedial measures to minimize pollution effects.

Reference

Schulze, E. D. May 19, 1989. Air pollution and forest decline in spruce (*Picea abies*) forest. *Science,* 776–83.

Case Study 30 AIR POLLUTION AND FOREST ECOLOGY

Name ____________________
Section ____________________
Date ____________________

I

Question 1—What other comparisons might have been made between trees at the 2 sites?

Answer 1

Question 2—Why was the photosynthetic rate examined in needles of varying age?

Answer 2

II

Question 1—What are the advantages in this experiment of using twigs on the same tree?

Answer 1

Question 2—Why is it necessary to check the effect of removing buds in *both* healthy trees and in trees showing evidence of damage?

Answer 2

III

Question 1—Why would air pollution damage not be as obvious if growth were slower?

Answer 1

Question 2—How would you explain Result 2 now after knowing Result 3?

Answer 2

Question 3—What other experiments could be performed to provide more information on nutrient uptake and availability?

Answer 3

31 Case Study: Co-Evolution (Difficulty Level 3)

In co-evolution, 2 unrelated groups of organisms act as selective agents on each other. While adaptations that are the result of selection pressures exerted by one group on the other seem obvious, experimental evidence to support this contention is often difficult to obtain.

One proposed example of co-evolution that has been extensively investigated is the relationship between the passionflower vine of the genus *Passiflora* and a butterfly, genus *Heliconius.* In experiments reported by Williams and Gilbert (1981), the butterfly lays eggs on the *Passiflora* plant and the larvae that hatch from the eggs feed on the young leaves and shoots. The *Passiflora* vines have evolved chemical defenses that are effective against most insects but not against *Heliconius.* Moreover, *Heliconius* deposits its eggs only on *Passiflora.*

An interesting observation is that some species of *Passiflora* have developed structures on the leaves and shoots that bear a striking resemblance to the bright yellow eggs of the butterfly. This observation has led to the following hypothesis:

Hypothesis—The leaf structures resembling eggs have evolved under selection pressure from *Heliconius.* The development of such structures is termed *egg mimicry.*

What possible advantage or protection would be conferred to the vine by the yellow leaf structures?

Prediction 1—The butterfly must have some characteristics such that the presence of an egg (or the appearance of an egg) on a leaf or shoot could inhibit the laying of additional eggs and thus provide protection to the vine.

Result 1—It is observed that the caterpillars of *Heliconius* are cannibalistic and will consume other caterpillars of the species that are nearby when they emerge from the eggs. Thus, there is apparent advantage to the butterfly to exhibit avoidance behavior (i.e., avoiding egg-laying sites where eggs are already present).

Is avoidance actually observed? If so, what cues are used by the butterfly to control the avoidance behavior?

A series of experiments and observations have investigated these questions. In one investigation, the number of eggs laid on individual vine shoots was examined.

If avoidance behavior is occurring, what would you predict?

Prediction 2—Vine shoots should have only one *Heliconius* egg laid on them.

Result 2—Observations show that shoots with only one egg are found more often than would be expected by chance. In other direct observations, *Heliconius* deposited eggs 70% of the time when visiting shoots with no eggs but only 30% of the time when visiting shoots where eggs were already present. Thus, there is an apparent preference by *Heliconius* to deposit an egg only on shoots where no eggs are present.

I What selection pressure is acting on the butterfly?
If such selection is effective in changing butterfly behavior, what would you expect this change to be?

While Result 2 does suggest that egg mimicry confers an advantage to *Passiflora,* a number of questions remain. What characteristic(s), such as color, shape, or odor, cues the butterfly to avoid sites where eggs are present? Are the mimic eggs of the plant as effective as real eggs in eliciting avoidance?

In another experiment, butterflies were given a choice of 3 egg-laying sites: shoots without eggs, shoots with eggs, and shoots with eggs that had been dyed green.

What would you predict would be the egg-laying behavior in this case?

Prediction 3—If egg color is the cue to avoidance behavior, then more eggs should be laid on unoccupied shoots and shoots with green-dyed eggs than on shoots with eggs.

Result 3—Eggs were deposited only 5% of the time on shoots with eggs compared to egg laying in 25% of the visits with both unoccupied shoots and shoots with green-dyed eggs. These results clearly suggest egg color is the cue that elicits avoidance behavior.

II In this experiment, the eggs on both the normal egg-bearing shoot and the eggs that were dyed green were thoroughly rinsed with distilled water before the experiment was started. Why was this done?
What was the purpose of including shoots with normal eggs?

While these results show that egg laying is far less prevalent on shoots that have eggs, egg deposition still occurred in 5% of the observations. Do the butterflies that lay eggs in spite of the presence of other eggs have no avoidance behavior or is there some inhibition even in these cases?

To examine this question, the time taken for the butterflies to lay eggs on egg-bearing shoots was compared to the time taken to lay eggs on shoots not bearing eggs.

What would you predict would be observed in this case?

Prediction 4—If avoidance behavior is expressed even when egg laying does occur on eggs-bearing shoots, then egg laying should take longer. In other words, the butterfly should hesitate for some longer period before actually laying an egg.

Result 4—Egg laying takes more than twice as long to accomplish on egg-bearing shoots (on the average about 22 seconds) as compared to egg laying on egg-free shoots (about 10 seconds). Thus, this experiment indicates that even when egg laying *does* occur in the presence of other eggs, there is some apparent avoidance behavior.

Although these results from the several experiments described are informative, evidence is needed that will test whether the mimic eggs on the vine (as opposed to the real eggs) are effective in deterring egg laying. Such a test was performed by comparing the frequency of egg laying on shoots bearing mimic eggs to egg laying on shoots that had mimic eggs removed.

Prediction 5—Egg laying should be more frequent on shoots where mimic eggs have been removed.

Result 5—Egg laying occurred on shoots with removed mimic eggs in 40% of the visits by *Heliconius* but in only 25% of the visits to shoots with mimic eggs. These results clearly indicate that the presence of mimic eggs inhibit egg laying by *Heliconius.* Since the presence of mimic eggs provides significant protection to *Passiflora* from egg laying and consequent caterpillar consumption of the plant, selection would favor the survival of those *Passiflora* plants that possess mimic eggs. Thus, the hypothesis that mimic eggs have evolved as a result of selection pressure by *Heliconius* is strongly supported by these experiments.

III Suppose egg laying occurred with equal frequency in Result 5. How might you interpret these results?
If a trial had been included where yellow dots (to simulate eggs) were placed on shoots, what would you predict would happen?

The series of experiments described here have tested several possible interactions, and each makes a distinct and significant contribution to testing the hypothesis. Testing a hypothesis, or testing the more detailed questions and predictions derived from that hypothesis, is necessarily a process of devising separate tests that will contribute to the ultimate interpretation.

Reference

Williams, K. S., and Gilbert, L. E. April 24, 1981. Insects as selective agents on plant vegetative morphology: Egg mimicry reduces egg laying by butterflies. *Science,* 467–69.

Case Study 31 CO-EVOLUTION

Name ____________________
Section ____________________
Date ____________________

I

Question 1—What selection pressure is acting on the butterfly?

Answer 1

Question 2—If such selection is effective in changing butterfly behavior, what would you expect this change to be?

Answer 2

II

Question 1—In this experiment, the eggs on both the normal egg-bearing shoot and the eggs that were dyed green were thoroughly rinsed with distilled water before the experiment was started. Why was this done?

Answer 1

Question 2—What was the purpose of including shoots with normal eggs?

Answer 2

III

Question 1—Suppose egg laying occurred with equal frequency in Result 5. How might you interpret these results?

Answer 1

Question 2—If a trial had been included where yellow dots (to simulate eggs) were placed on shoots, what would you predict would happen?

Answer 2

32 Case Study: Phylogeny (Difficulty Level 3)

A variety of techniques and lines of evidence have been used to investigate and establish the evolutionary relationships among organisms. One of the more recently developed techniques relies on changes that occur with time in an organism's DNA. During the process of evolution, as new species diverge from common ancestors, genetic differences accumulate through selection and nucleotide substitutions.

It is logical then to hypothesize the following:

Hypothesis—An accurate measure of how closely related classes of organisms are to each other can be made by comparisons of the DNA of these organisms. The magnitude of any difference between the organisms would be a measure of the time since the species or genera began evolving from a common ancestor.

One possible comparison of DNAs would be to determine the sequence of nucleotides in each DNA and compare the differences. Such a process, however, would be unreasonably time-consuming. Recently, another comparison has been developed that is based on the fact that double-stranded DNA is comprised of 2 single strands held together by hydrogen bonding between complementary base pairs. If a solution of double-stranded DNA is heated, the relatively weak hydrogen bonds are broken and the 2 strands separate. Now, if this solution is cooled, what would you expect to happen?

Prediction 1—Hydrogen bonds will reform between complementary bases, and the double-stranded molecule will be regenerated.

Result 1—Observations indicate that the original double-stranded molecule is regenerated. The cooler temperature allows the hydrogen bonds to form and remain stable as complementary bases come into contact. Actually, for hydrogen bonding to be stable, large segments of the 2 strands must be matched appropriately along the length of the strands. This means that a few chance associations between individual complementary bases will not form a stable 2-stranded molecule.

I Would you expect different amounts of hydrogen bonding between DNA from 2 individuals of the same species? Why or why not?

Do you think it would be possible to find 2 species that appear very similar but whose DNA exhibits less hydrogen bonding than DNA from 2 species that appear very different? Why or why not?

What if single strands of DNA from different but apparently related species were mixed in a heated solution and this solution were allowed to cool? What would you predict?

Prediction 2—Segments of the 2 strands that are complementary will form hydrogen bonds. The greater the number of hydrogen bonds, the more similar the species.

Result 2—Further measurements have shown that there is a greater degree of recombining (i.e., hydrogen bonding) between single-stranded DNA from 2 species that are closely related than between DNA strands from more distantly related species. The degree of hydrogen bonding is measured by the temperature at which the bonded DNA strands from different species (termed *hybrid DNA*) separate. The greater the number of hydrogen bonds, the greater the temperature at which they separate.

II What if the rate of DNA change were not constant? How would this affect any conclusions that might be made?
What if the rate of change in the DNA from an organism subject to intense natural selection were compared to the rate of change in an organism experiencing relatively little selection pressure? Would you expect any differences?

The evolutionary relationships between species that were utilized in Result 2 to compare DNA recombining data were obtained from the fossil record. In a great many cases this is not possible, however, and DNA hybridization comparisons are used in obtaining information on possible evolutionary relationships. So far these results give only a qualitative sense of the evolutionary relationships. A measure of the *time* since 2 similar species have diverged from a common ancestor would be far more useful. Can the amount of hydrogen bonding (i.e., the degree of hybridization) be used to measure time since divergence?

Investigators have devoted considerable effort in developing a "molecular clock" whereby the amount of change in some molecule, such as DNA, can be used as an indicator of evolutionary time. A critical question is: Do changes (in the case of DNA, the number of changes in the nucleotide bases making up the molecule) occur at a regular rate over time or are these changes variable? Sibley and Ahlquist (1983) describe extensive investigations of bird phylogeny, including numerous tests of the rate of DNA change. Using sets of 3 bird species, 2 of which were more closely related to each other than to the third, DNA hybridization comparisons were made between each of the closely related species and the more distantly related species. In each case, the 3 species shared a common ancestor and thus the 2 more closely related species had been separated from the more distantly related species the same amount of time. This procedure is shown in figure 32.1.

Figure 32.1 Species *x* and *y* have evolved independently since branching from the common ancestor *a*. If the rate of DNA change is constant, then the same amount of DNA change should have occurred between *a* and *y* as occurred between *a* and *x*. This can be tested by DNA hybridizations between species *x* and *z* and between *y* and *z*. If the same amount of change is found in the 2 hybridizations, this result suggests that the molecular clock runs at a constant rate.

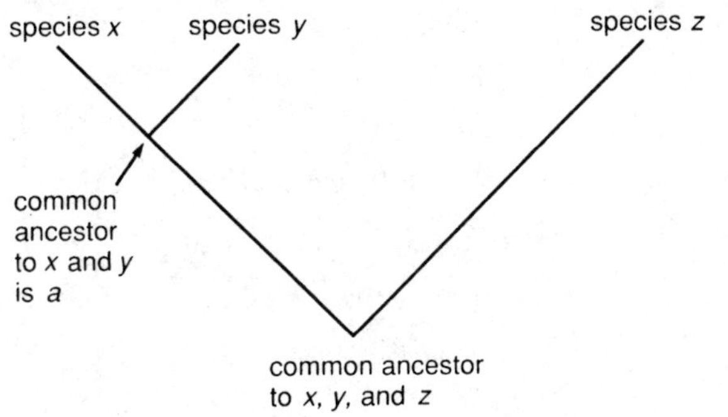

What would you predict would be found when the amount of DNA change is compared in these species?
Prediction 3—If the rate of DNA change is constant (i.e., if the molecular clock runs at a constant rate), DNA hybridization should show an equal amount of difference between each of the closely related species and the more distantly related species.

Result 3—The predicted results were observed in literally thousands of comparisons within sets of 3 related bird species. These results provide strong support for a molecular clock that runs at a constant rate. This may not be the case with all classes of organisms, however, and is the subject of much debate among investigators.

DNA hybridization experiments performed between ostriches (Africa) and rheas (South America) provide an example of the way the dating of evolutionary time is accomplished. At one time the common ancestor of these 2 birds roamed a large landmass that later split apart to form Africa and South America. Seafloor spreading and continental drift led to increasingly greater separation of these continents. Geological data indicates that the continents split apart about 80 million years ago. At this time, the 2 populations of birds in Africa and South America became separated and began to evolve independently. Knowing this time of separation and the amount of DNA change from hybridization establishes the DNA change per unit time and provides a time scale for measuring evolutionary time for other bird species.

III Suppose the birds under investigation in the example were capable of flight instead of being flightless like the ostrich and rhea. Would this have affected the established time scale?

As a specific example of how DNA hybridization studies can yield new data or resolve controversy, consider the investigations of the relationship between the new world vultures and old world vultures. While these 2 groups of vultures appear and behave similarly, it was argued on the basis of certain morphological characteristics that new world vultures, including condors, are more closely related to storks. Subsequent DNA hybridization investigations confirmed that storks and new world vultures are most closely related to each other. Apparently the similarities between new and old world vultures are the result of convergent evolution.

The technique of DNA hybridization in combination with the careful study of geological events and fossil evidence has allowed new investigations of phylogenetic relationships. The experiments described here provide strong support for the hypothesis that changes in DNA can be used to establish evolutionary relationships and evolutionary dating.

Reference

Sibley, C. G., and Ahlquist, J. E. 1983. The phylogeny and classification of birds based on the data of DNA-DNA hybridization. *Current Ornithology,* 245–92.

Case Study 32 PHYLOGENY

Name ____________________
Section ____________________
Date ____________________

I

Question 1—Would you expect different amounts of hydrogen bonding between DNA from 2 individuals of the same species? Why or why not?

Answer 1

Question 2—Do you think it would be possible to find 2 species that appear very similar but whose DNA exhibits less hydrogen bonding than DNA from 2 species that appear very different? Why or why not?

Answer 2

II

Question 1—What if the rate of DNA change were not constant? How would this affect any conclusions that might be made?

Answer 1

Question 2—What if the rate of change in the DNA from an organism subject to intense natural selection were compared to the rate of change in an organism experiencing relatively little selection pressure? Would you expect any differences?

Answer 2

III

Question 1—Suppose the birds under investigation in the example were capable of flight instead of being flightless like the ostrich and rhea. Would this have affected the established time scale?

Answer 1

33 Case Study: Mating and Reproduction in Jackals (Difficulty Level 3)

Jackals are members of the family Canidae, which includes the foxes, dogs, and wolves. Extensive observations by Moehlman (1987) have shown that jackals often mate for life, a characteristic unusual among mammals. In addition, some offspring frequently remain with the parents to help care for the next year's offspring.

What accounts for this relatively unusual social organization of jackals? A major hypothesis is as follows:

Hypothesis—Such behavior favors the survival of offspring within the particular environment occupied by these animals.

Testing this hypothesis has involved several years of observation and data collection on 2 species in the plains areas of Africa—the silverback jackal and the golden jackal. Although both species are omnivores and feed on a variety of food sources, the primary food sources of the 2 are different. Silverback jackals feed mainly on small rodents and the datelike fruit of the *balanites* tree. These feeding habits require that silverbacks forage widely for food. Golden jackals, on the other hand, occupy smaller territories and rely mainly on the meat of large vertebrates.

The family Canidae show a trend toward larger litters as the body size of the species increases and jackals are consistent with this trend. Large litters of pups, generally 6 per litter, are produced by both species. Larger litters, however, result in smaller and less developed pups at birth.

Do these characteristics suggest reasons for the unusual social organization of jackals? One possible consequence of large litters of less developed pups would be that rearing these pups would require more extensive parental effort due to extended infant dependency over a longer period of time until these young can care for themselves.

What would this observation suggest about the required involvement of both the male and female parent in rearing the young?

Prediction 1—Species bearing larger and more dependent litters will exhibit more long-term and enduring mating between parents (i.e., pair formation). In other words, these species will have a stronger tendency to mate for life because the efforts of both parents are needed to care for the young.

Result 1—Silverback jackal parents manage to rear successfully only an average of 1.3 pups per litter from a litter size of 6. The male parent is a full participant in the care of the pups; if this effort were not invested, probably *no* pups would survive. Long-term pair-bonds are apparently necessary to provide the required infant care.

Golden jackal care by both parents does not appear to be as critical since food procurement is not as difficult. Greater success is observed in raising pups with an average of 1.8 surviving. It is also observed that pair-bonding is not as enduring in golden jackals.

These results thus support the prediction and indicate why enduring pair-formation is a significant advantage to the survival of young.

I Why is food procurement more difficult for silverback jackals?
In addition to difficulty in obtaining sufficient food for the offspring, in what other ways would food procurement create problems in rearing the pups of silverbacks?

Do previous offspring that remain with the parents and serve as helpers with infant care provide a similar advantage?

Prediction 2—Helpers remaining with parents should result in the survival of a greater number of offspring.

Result 2—As shown in figure 33.1, the presence of helpers significantly increases the survival of pups in both silverback and golden jackals.

II The data shown in figure 33.1 suggest that helpers provide a greater survival advantage for silverback pups than for golden jackal pups. Why should this be so?

Figure 33.1 Number of pups surviving per litter as a function of the number of adults per family (parents + helpers).

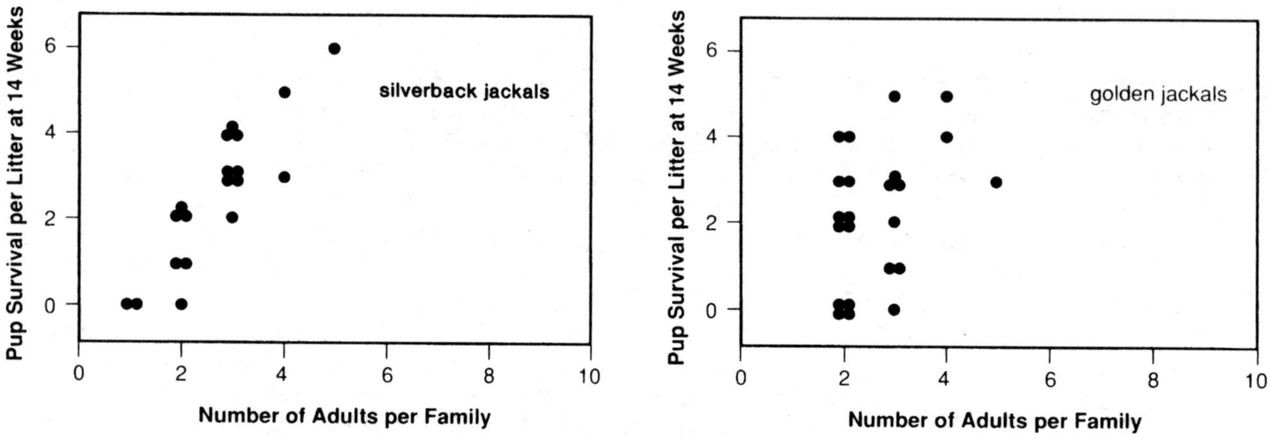

In what ways can helpers provide an advantage in raising pups? Some possibilities should be immediately obvious. Assistance in obtaining food for the young would be an immediate and necessary benefit. Added help in protecting the young would be another benefit.
Prediction 3—Helpers should be observed assisting with feeding and protecting the young.
Result 3—Helpers have been observed contributing up to 33% of the pups' food. In families with 2 helpers, pups are left alone only 8% of the time, but in families with no helpers, pups are left alone as much as 20% of the time. These results clearly indicate that helpers make highly significant contributions to both food collection and protection of the young.

III Female helpers do not reproduce even though they are sexually mature. Is this lack of reproduction a benefit to the family?
After successful hunting and eating, jackals typically return to their dens and regurgitate food for the young. Why is this a successful strategy for providing food for the pups?

Is there any benefit to the helpers in staying and helping with the pups? Why should a helper stay with the family rather than migrate to a different territory and establish a new family?
These questions have led to the following prediction:
Prediction 4—Helpers should derive identifiable benefits from staying with the family and assisting with raising the young.
Result 4—One major reason why staying may be beneficial to helpers is that pups are full siblings (i.e., brothers and sisters) of the helpers and they share common genes. By helping to ensure survival of the pups, helpers are increasing the probability that their genes will be passed on to future generations. Thus, there is a significant advantage to the evolution of "helping" behavior.
Other observations reveal differences in environmental factors, which in turn may affect dispersal of offspring. In silverbacks, food is relatively abundant (although dispersed), and only 24%

stay as helpers. It is difficult though, for migrating young to raise pups, and staying with the family another year might provide young silverbacks with experience in foraging for food and rearing young.

In golden jackals, there can be serious difficulties in acquiring a territory for a new family due to a high population density of golden jackals. Failure to acquire a mate and territory apparently results in the return of 70% of the young jackals to the family for a period, during which time they serve as helpers.

Although considerably different, these 2 environmental situations are instructive because the end result is the same in both jackal species—young adults remain with the family for a longer period of time as helpers.

These results show that careful observational studies can suggest ways relatively complex behavior can evolve and develop in response to a variety of pressures. A number of environmental and evolutionary factors were considered and contributed to an understanding of the social organization of jackals.

Reference

Moehlman, P. D. July–August 1987. Social organization in jackals. *American Scientist,* 366–75.

Case Study 33 MATING AND REPRODUCTION IN JACKALS

Name ____________________
Section ____________________
Date ____________________

I

Question 1—Why is food procurement more difficult for silverback jackals?

Answer 1

Question 2—In addition to difficulty in obtaining sufficient food for the offspring, in what other ways would food procurement create problems in rearing the pups of silverbacks?

Answer 2

II

Question 1—The data shown in figure 33.1 suggest that helpers provide a greater survival advantage for silverback pups than for golden jackal pups. Why should this be so?

Answer 1

III

Question 1—Female helpers do not reproduce even though they are sexually mature. Is this lack of reproduction a benefit to the family?

Answer 1

Question 2—After successful hunting and eating, jackals typically return to their dens and regurgitate food for the young. Why is this a successful strategy for providing food for the pups?

Answer 2

34 Case Study: Balanced Polymorphism (Difficulty Level 3)

Balanced polymorphism is the maintenance of genetic variation due to environmental pressures. For example, in sickle-cell disease, the allele frequency for Hb^S and Hb^A is maintained in African populations because of heterozygote superiority in the presence of the malaria parasite.

This case study considers an example of polymorphism in the mussel *Mytilus edulis* off the shores of Long Island. To the south of the island there is the Atlantic Ocean and to the north there is Long Island sound, which separates Long Island from the mainland. The ocean waters are much more salty than those of the sound because the sound receives freshwater runoff from the mainland. We will be considering 2 populations of *M. edulis*—one that lives in the ocean and one that lives in the sound.

Would you expect the allele frequencies to vary between the 2 populations? Why?

The cells of *Mytilus edulis* contain a type of enzyme called aminopeptidase I, which occurs in 3 variations distinguishable one from the other by certain physical properties. For each of these aminopeptidase I enzymes there is different allele, designated as *lap* 94, *lap* 96, and *lap* 98. Considering our discussion so far, would you expect to find a different frequency, for instance, for *lap* 94 in the 2 populations?

Prediction 1—*Lap* 94 frequency should be different in the 2 populations because the environment varies for the 2 populations.

Result 1—Measurements indicate that the frequency of *lap* 94 allele is 55% in the population on the ocean side of the island but only 12% for the population in Long Island sound.

I What is the sum frequency of *lap* 96 and 98 in the ocean population? Which of the alleles (*lap* 94 versus 96 and 98 taken together) seems to confer an advantage to mussels living in the ocean? What is the sum frequency of *lap* 96 and 98 in the sound population? Which of the alleles (*lap* 94 versus 96 and 98 taken together) seems to confer an advantage to mussels living in the sound?

Mussel larvae (but not the adults, which are permanently attached to a substrate) are mobile, and it has been noted that mussel larvae from the ocean migrate each year into the sound on prevailing ocean currents. Despite migration of these larvae, the frequency of *lap* 94 stays the same in the sound population.

Why might you have assumed that the frequency of *lap* 94 would continually rise in the sound population? What explanation can you give for the frequency not rising? On the other hand, why wouldn't the allele be completely eliminated?

The much higher frequency of *lap* 94 in the ocean population suggests that the *lap* 94 allele confers an advantage to mussels living in the ocean. What is the possible advantage? Additional research into the role of aminopeptidase I, an enzyme that breaks down polypeptides to amino acids in mussel cells, helps answer this question. In many marine organisms, intracellular free amino acids osmotically balance the salts in seawater. A cell with a low content of amino acids will tend to become dehydrated, but a cell with a higher content of amino acids will not become dehydrated. Would you expect greater aminopeptidase I activity in mussels exposed to seawater of high salinity?

Prediction 2—Aminopeptidase I activity should be greater when mussels are in seawater of high salinity.

Result 2—Experimental results have shown that aminopeptidase I activity is greater in mussels exposed to seawater of high salinity than when they are exposed to seawater of low salinity.

Aminopeptidase I activity apparently helps a mussel respond to an increase in the osmolarity of the environment.

In regard to the *lap* alleles, *lap* 94 codes for an aminopeptidase I enzyme that is more active in breaking down polypeptides than the other 2 alleles. Mussels having which type of allele do you predict would be better able to respond to a high-salinity environment by producing amino acids?
Prediction 3—Mussels with the *lap* 94 allele should be able to respond quickly to a high-salinity environment by producing free amino acids because the aminopeptidase I enzyme coded for by this allele exhibits greater activity.
Result 3—Subsequent measurements reveal that this prediction is valid. Mussels with the *lap* 94 allele are able to control cellular osmolarity in waters of high salinity.

II Summarize the data given in Results 2 and 3. Why does the presence of the *lap* 94 allele help mussels "control cellular osmolarity in waters of high salinity"?

The presence of the *lap* 94 allele provides a distinct advantage in mussels living in high-salinity waters. If mussels adjusted to a high-salinity environment are placed in a low-salinity environment, however, they excrete waste products characteristic of amino acid breakdown. The loss of amino acids prevents their cells from taking in too much water. Would you expect mussels with the *lap* 94 allele to be at a disadvantage, however, when placed in a habitat that changes in salinity from high to low and back again?
Prediction 4—In some environmental conditions, such as one that has frequent changes in salinity, mussels with the *lap* 94 allele would be at a disadvantage.
Result 4—Investigations of nitrogen balance in *M. edulis* indicate that mussels with the *lap* 94 allele suffer more nitrogen loss than mussels with only alleles *lap* 96 and/or 98 in an environment like the entrance to Long Island sound, where salinity changes are great. When the salinity rises, they can quickly adjust, but when it lowers, they lose more nitrogen than otherwise. When nitrogen loss is sufficiently great, mussels cannot retain sufficient amounts of this nutrient to survive.

III Suppose the environment could provide nitrogen to make up for any lost by the mussels. Would you expect the frequency of the *lap* 94 allele to vary significantly between the 2 populations? Why not? Since the frequency of the *lap* 94 allele does vary between the 2 populations, what does this say about the availability of nitrogen in their diets?

These experiments and observations strongly support the hypothesis that genetic variation is maintained by selection pressures acting on the *lap* alleles in different environments. A remaining question, however, is why should both the *lap* 96 and *lap* 98 alleles persist in the population? Both are apparently characterized by a lower rate of polypeptide breakdown. Is selection on some other property acting to maintain both alleles? Further studies will be necessary to investigate additional physiological roles of these 2 alleles.

Reference

"The Adaptive Importance of Genetic Variation," by R. K. Cohen and T. J. Hilbish, *American Scientist,* vol. 75, pp. 134–141 (March–April 1987).
Cohen, R. K., and Hilbish, T. J. March–April 1987. The adaptive importance of genetic variation. *American Scientist* 75:134–41.

Case Study 34 BALANCED POLYMORPHISM

Name ______________________
Section ______________________
Date ______________________

I

Question 1—What is the sum frequency of lap 96 and 98 in the ocean population? Which of the alleles (lap 94 versus 96 and 98 taken together) seems to confer an advantage to mussels living in the ocean?

Answer 1

Question 2—What is the sum frequency of lap 96 and lap 98 in the sound population? Which of the alleles (lap 94 versus 96 and 98 taken together) seems to confer an advantage to mussels living in the sound?

Answer 2

II

Question 1—Summarize the data given in Results 2 and 3. Why does the presence of the lap 94 allele help mussels "control cellular osmolarity in waters of high salinity"?

Answer 1

III

Question 1—Suppose the environment could provide nitrogen to make up for any lost by the mussels. Would you expect the frequency of the lap 94 allele to vary significantly between the 2 populations? Why not?

Answer 1

Question 2—Since the frequency of the lap 94 allele does vary between the 2 subpopulations, what does this say about the availability of nitrogen in their diets?

Answer 2